...Not

Fade Away

Written by the Veterans and friends of
the Rochester Veterans Writing Group
Rochester, New York

RVWG
VERITAS PER VIRTUTEM
MMXIV

These beautifully written, tender, heartwarming, and honest accounts by the Rochester, NY Veterans' Writing Group, must be shared to help civilians understand what the people who serve in the military sacrifice for us and for our freedom. These poignant vignettes will warm your heart and open your eyes. The first pages shed light on what it is like to be a military wife or family member of someone who serves. And the following writings, penned by these honorable veterans will remain in your heart. I promise their words will Not Fade Away.

- Linda Mazur, *President: The Emilee Connection*, Author of *Emilee: The Story of a Girl and Her Family Hijacked by Anorexia*

...NOT

Fade Away

The old barracks ballad declares:
"Old soldiers never die, they just fade away."

But it was Old Soldiers
who served, suffered and sacrificed
so the rest of us could remain free.

The Rochester Veterans Writing Group

Offers you these veterans' stories
in the belief that old soldiers should

...Not Fade Away

...*Not* Fade Away

APRIL 1968 and "WELCOME HOME" appear as Chapters 14 and 15, respectively, in *Once a Boy*, by Paul Gardner, Mystic Dragon Publishing, Baldwin Park, CA ©2021

ISBN 978098417922-0

Cover Art by Gary Redlinski

Edited by Green Shadow press

Contents

Sue Spitulnik

Sue Spitulnik was an Air Force wife, living in four states and England from 1972 to 1979, and worked until 1990 in a Tacoma, Washington restaurant where the clientele was military connected. She moved back to New York, her home state, in 1991.

In 2015, Sue attended a meeting of the Rochester Veterans Writing Group in search of an Air Force pilot to interview for a writing project. She remains active in the group and has a memoir essay published in the group's first book, ***United in Service, United in Sacrifice***. Sue also creates 99-word flash fiction stories based on a weekly prompt presented at Carrot Ranch Literary, an international online writing group. She has added each scene to a serial entitled "Michael's Circle" about an Army double-leg amputee and his close family and friends for three years on her blog at suespitulnik.com.

Sue is the current president of Lilac City Rochester Writers, a group that focuses on the craft of writing and publishing for fiction writers. She has a short story published in their ***No Page Left Unturned*** anthology. Both books are available on Amazon.

Fairport is where Sue now lives with her husband, Bob, and two cats. She is quilting when she isn't writing, specializing in patriotic and t-shirt quilts.

My Veteran Friends

By Sue Spitulnik

These are some of the most memorable stories shared by my personal veteran friends in the Rochester Veterans Writing Group.

One man tells a story about waiting to find some alone time with his wife to tell her he was going on deployment again, this time to Afghanistan. He finally found it near their washing machine while she folded clothes. When he told her, out of earshot of their children, she stared at him and asked, "Where do you want to be buried?" Thankfully he returned home, and they now have two grandchildren. I owe a debt of thanks to him, for had he not started the group, I would not have the extraordinary friends mentioned here.

I loved hearing the reaction of two vets when they realized through their storytelling that they had both helped themselves, along with other soldiers, to the sausages in a shop in Europe during World War II. In both cases, the shop owner stood back and watched. Both men said, "We meant the owner no physical harm; we were hungry." After the war, the sausage maker moved to Buffalo, NY, and both vets, at separate times, visited him. They apologized for their actions during the war and shook hands, agreeing that forgiveness was better than holding a grudge. Sadly, these two veterans are no longer with us. The sausage story gets retold regularly, accompanied by much laughter and moist eyes.

My favorite story told by the older WWII vet was about when he and a buddy used depth charges in a small lake to shock the fish, so they came floating to the top. It's not a legal way to fish, but their unit had fresh food for supper, and no one asked who procured them. This same man later in life made pets of the bullfrogs in his pond, could tell their croaks apart, and called them out of the water to get treats. He also became a world-known abstract artist. He has pieces on permanent display in multiple museums, and I own and cherish one of his multi-media 3-D owl pictures.

By only a few months, the younger of the WWII vets told us stories about repeatedly getting separated from his unit because he went exploring when he shouldn't have. On one such occasion, he showed his weapon to commandeer a motorcycle from an enemy soldier, left the man standing on the road, and rode off to get back to his unit. It was hard for me to picture the refined gentleman whom I spent time with when he was in his nineties as a young cut-up. Both of these men's photos have a place of honor on my refrigerator door. I miss their spark.

I admire someone who has one of his paintings on permanent display in the George W. Bush Presidential Library. I'd say that's a fine accomplishment. He made a dream of the younger WWII vet mentioned above, to retrace his war-time steps in Europe before passing, come to fruition by traveling with him to do just that. They managed to locate and talk to some families who shared the same memories. Today he is active in helping other veterans cope with everyday life after the military. He and his companion dog are always welcome at any event.

A woman I am fond of told about a time she had gathered in New England for a fun vacation with her parents, siblings, and families. After only one night together, her active-duty brother got a call from his commanding officer about an emergency in his unit. He packed up his family immediately and headed back to Florida. So much for family time. His children and their cousins were not happy. Military-connected folks know and accept, that this can happen at any time to a service member.

I'm close to a more than average soldier who asked for thermodynamics books to read in the jungle for his own entertainment. Really? Most of us would find his choice to be a burden. He is also known to have taken coffee to lookout tower guards in the dark of night, so they knew they weren't alone. And to me, he delivers flowers on occasion. He's a special guy who can make anyone feel important when he talks to them.

Another man who has a special place in my heart was assigned to the tent morgue in Vietnam. He says that he didn't feel he had contributed enough to his "brothers" after one tour, so he signed on for another. The consensus of our group is that none of us would have been able to perform his duties. He relates that his whole person was permeated with the smell of death to the point no other soldiers who didn't work in the tent would eat near him and his cohorts. He would laugh and say, "What smell?" It's called morgue humor. He admitted that he was having difficulty dispelling some memories made prominent during a writing session, so he has turned to painting and wood burning for creativity. I know where his art studio is and visit him there. I have two pieces of his artwork at home and appreciate their value beyond the dollar amount.

While still on active duty and in the hospital, one soldier had a "you can go home next week" turn into many more months of care and waiting for the proper paperwork to be issued. During that extended time, his wife, who lived on the other side of the country, decided she had waited long enough and left him. Another story he told about a yellow flower outside the "perimeter fence" stays with me. He and his fellow soldiers saw it when they were stationed where everything was the same color, sand. The bright yellow and sickly green assured them there was life on the outside. He also tells of hearing the blasting call-to-prayer speakers in the nearby town that warned of a nighttime mortar attack on their installation. He laughs at the irony of being warned.

One of the other military wives in the group has become like a sister. She kept the home front together for years for her often-irritable Korean war veteran husband. When their grandchildren were born, she witnessed the blossoming of a very gentle man she didn't expect to appear. She says it was worth the wait.

A man I have grown close enough to, so I tell others he is my chosen older brother was a pilot of a C-130 airplane in the Air Force. While in the service, he had to land the giant airplane on an icy runway in Iceland. Because of the conditions, they came in sideways. He was able to get the plane straightened out, and neither the aircraft nor the contents were damaged. I don't know about the underwear worn by those involved. He shared another experience: while in training, he and his crew witnessed a training-flight plane crash near them while walking out to their aircraft. They held their breath until they sighted the endangered team walk out of the flames, then had to fly their own training mission minutes later. That takes resolve, guts, and a whole lot of concentration.

I am honored to have become friends with an African-American lady who is an RN. When she was a young WAF, an older white officer poked her in the chest and said, "You'll never amount to anything!" I hope that no longer happens in our military. I also hope that man knows she advanced through the ranks with her determination and hard work and retired as a Lt. Colonel. One of her final assignments was to identify body parts and place them in the correct body bags to return to the US and their families. The experience haunts her.

I've also had the privilege of getting to know a lady who helped start the Rochester Veteran's Outreach Center many years ago. She and some other local vets, one a Medal of Honor recipient, focused on helping service members with PTSD. She's never quite gotten over the fact that the bean counters stepped in and changed the organization's focus to helping vets with housing and job needs because there's money in it. We know if PTSD is dealt with first, gaining the other necessities of living will automatically follow.

I have had to say a fnal goodbye to an Army soldier who told us about singing the same songs repeatedly to keep himself awake while on guard duty. He helped us have a good conversation about stemming boredom. When he provided the prompts for our group, we knew we would have a thought-provoking discussion. We miss his unassuming ways, neat notebook, and how he connected with others.

I'm also friends with a woman who experienced the upper-class atmosphere of being a military doctor's wife, and now her son is a Marine officer. Both types of home-front warriors are a privilege and too often stressful, depending on the day. She currently works counseling other young veterans and can understand them better because of her own experiences. She tells of being a member of the Officer's Wives Club who made it their mission to be on the flight line any time a plane landed carrying the casket of a fallen soldier. They believe anyone who has paid the price for our freedom should be welcomed home by people who understand.

Another special man is a family member of a WWII Army doctor. He is consolidating the hundreds of his father's letters for all to read. It will be an accurate picture of life during WWII for a service doctor and the family left at home for three years with no visits. There aren't a lot of those stories on record. I'll look forward to having a copy of his book for my descendants so they can know about that reality.

Oh, and I can't leave out my husband. Way back when while in the Army, after digging latrines all day, he took a tank to drive back to base so he could get a hot shower. The story goes that he didn't get into trouble because his Commanding Officer owed him money from a few poker games. He's still an excellent card player.

And if my friends were to tell you something about me, they would say that I am a good organizer and add that I opened my home during the pandemic while meeting rooms were closed to groups like us. But they would be thinking, ***she cries.*** Yes, I wear my emotions on my sleeve, and I cry at almost every gathering because I feel everyone's hidden pain.

From recounting my strongest memories of what my friends have shared, I hope you now understand how true the slogan "Freedom Isn't Free" is. Military members have all paid with their service and sacrifice, and so have their families.

The Cost of Freedom

By Sue Spitulnik

Have you ever sat around the dinner table with family discussing how much it will cost to fly across the country or perhaps overseas to visit one of your children who is stationed there with the United States military? Have you marked the weeks and months off a calendar before that child, actually an adult, will come home on leave or return to a close stateside installation because you can't afford the trip?

As a spouse, have you ever had your mate thrash in bed, wake up in a soaking sweat, and not know where they are? Have you had to calm your spouse on such an occasion hoping they quickly realize it was a dream, perhaps a recurring horrible dream, but they are safe, physically? Have you had the privilege of going to a counseling session with your spouse to attempt to eradicate that dream once and for all? Privilege? Yes, that means they are getting some much-needed help, having first been able to admit they need it.

As a child of a military person have you had to say goodbye knowing you won't see that parent for close to a year because they are leaving on an unaccompanied tour? Or, have you been expected to call them by their rank instead of Dad or Mom? Have you changed schools and addresses so often you have no one place to call home, nor a close relationship with any relatives? Or as a spouse, have you experienced these things and given up your roots to be a home-front warrior.

As a family member at home have you ever not been able to sleep because you are fretting your child/spouse is in harm's way in some foreign country while they serve? Have you been afraid to answer the telephone after you've heard of escalating fighting in said country? Or have you feared seeing two men in military uniform coming to your door to tell you your beloved is missing or dead?

Have you ever heard someone say, "All he had to do was escort a casket home. Why should that be a problem?" and want to slug the speaker in the head? Because as a person connected to the military you know the person in the casket gave his all and so did his/her family. Have you heard someone say only people who actually fought in the front lines of a war should be called veterans with no understanding that it takes twenty to twenty-five people in the background to keep one soldier armed, clothed, fed, sheltered, and paid?

And in my opinion, have you heard the worst, most ignorant comment of all, "They knew what they signed up for?" More than likely, they had no idea they would immediately be taught to think as a member of a team, not as an individual, and that the experiences they have and the traveling they do broadens their perspective, so it's hard to fit in back home when they leave the military. They don't realize that even if the service member gets letters and, today, face-time calls, they don't know the day-to-day happenings of their core family and neighborhood friends, so it's hard to join the conversation when they visit the family. And more than likely, the job they do in the military won't be transferable to a good-paying job once out of the service.

Am I bad-mouthing the armed forces' experience? Absolutely not. I'm pointing out common happenings in the military that non-military families don't know about and never thought of having to deal with. It's a statistical fact that less than 10% of all U.S. citizens have served in the military. It's a fact that often it is many members from the same family line for generations that serve. The United States is a big country, it's probably illogical to expect everyone to serve as is required in some small countries, but that doesn't change the fact that so few pay for the freedom of so many. The slogan "freedom is not free" is very true. The armed forces members who signed up to pay with their lives if necessary and their families who are affected understand the cost.

I was an Air Force wife from 1972 until 1979. I experienced a lot of the examples above, and if I didn't, I know someone personally who has. I enjoyed the Air Force life of living in different places, including England, meeting new people, having a dependent military ID that gave me the right to go on military installations without question to shop and play at reduced rates, and having access to medical care of any kind when my family or I needed it. If our veterans had the same easy access to medical care as they were promised—the type that active-duty forces enjoy—it would be a grand thing. Unfortunately, it is an unkept promise to many veterans.

My Unsung Heroes

By Sue Spitulnik

A friend of mine posted this on Facebook at Christmas time a few years ago. He admitted he had copied it from someone else's page. It spoke to me, so I copied it and am repeating it as a testament to my unsung heroes, the combat veterans, those who it takes to support them, and those left at home carrying on like they aren't terrified of seeing two men in military uniform come walking up their sidewalk. It was written about Christmas but could just as easily apply to any big family event.

Quote:
If you have a Combat Veteran in your family and you don't like their moods and behavior around the holidays; please consider these six things:

1.) Your combat veteran has served in countries where people are blessed to receive a tattered pair of shoes or have clean water to drink; they no longer live the "first world illusion" and no longer care that if you buy one Play Station you can get a second one for fifty percent off. In fact, they find it hard to appreciate any of the gluttonous commercialism and overindulgence that permeates American holidays. Standing watch, boring as it was, had so much more purpose than going to the mall.

2.) Your Combat Veteran is thankful for the most basic things; not thankful for mega-sales and million-dollar parades. They are thankful to be alive; thankful to have survived both the wars far away and the wars they struggle with inside.

3.) Your Combat Veteran is thankful that it wasn't them that got killed, but their celebrations are forever complicated by guilt and loss over those who were. Some of the most thankful times in their life were some of the scariest. Their feelings of thanks and celebration often conjure memories that are equally painful.

4.) Your Combat Veteran is not like you anymore. At some point, for some period of time, their entire life boiled down to just three simple things:

a. When will I eat today?
b. When will I sleep today?
c. Who will I have to kill or will try to kill me today?

They are not like you anymore.

5.) Your Combat Veteran does not need a guilt trip or a lecture; they already feel detached in their grief while others so easily embrace the joy of the season. They need understanding and space; empathy not sympathy.

6.) Your Combat Veteran does love their family and is thankful for the many blessings in their life...and they are thankful for you.

To all my brothers and sisters of the uniform, know that we all struggle with one thing or another but as we go into this holiday season, reach out to those you love. You didn't fight alone on the battlefield and you don't have to fight alone at home. Unquote.

I know enough combat veterans to verify these six things are true, and I know even more non-combat veterans that still deal with all but number four on this list. When my sister died in 2020, we found a stack of letters from her college boyfriend, Wayne B. He enlisted in the Air Force in 1966 and went to Vietnam but was stationed way back from the front lines as an airplane mechanic. He wrote of his guilt for having a "safe" job and not being in harm's way with his buddies. He didn't earn a Purple Heart for a physical wound, but I can attest that his psyche was wounded. The wedding he and my sister discussed before he left the states did not happen when he returned, nor ever.

I've known veterans from Vietnam and subsequent wars who have returned home in the flesh but have difficulty fitting in because of their experiences. I describe them as "Prisoners of War," walking our streets, working jobs, and often drinking to excess because the military reshapes their thinking to be a part of a team, and they no longer have a team or tribe, so they flounder. Many figure out a way to get on with their lives, but others become a statistic.

It was recently related to me by a combat veteran that often when you see a person wearing the hat of their unit of Armed Forces, it serves a two-fold purpose. They are proud to be a veteran and would serve again if necessary, but also to identify themselves to other veterans, their brothers and sisters who understand their willing sacrifice.

I applaud them. They are all heroes to me. They should all receive Medals of Honor, but I know that is unrealistic. I am thankful for all of them.

Cultural Differences in England

By Sue Spitulnik

English as a second language for an American doesn't seem like an accurate statement unless you have lived in both countries, the United States first and Great Britain second.

In 1972 I became an Air Force wife. I felt I was entering the excitement of marriage and added adventure, knowing I would be traveling with my husband. We experienced four states and three years in the United Kingdom in seven and a half years. It was in Britain that I learned I spoke American, and they spoke English with a heavy accent, depending on where in the country one lived. It took me six months to be able to eavesdrop on a conversation and understand what was being said.

The first word I learned a new meaning for was probably fag. In Britain, it means cigarette, though I may have known that from movies about WWII. My husband at the time liked to go to pubs, and when we could get a babysitter for our son, or I wasn't pregnant with our daughter, I went with him. Smoking was allowed inside in the '70s, and sharing a pack of smokes was common. Marlboros were the favorite brand of the Brits, and that is what he smoked. Because English cigarettes were expensive, swapping a couple of American, inexpensive military-bought smokes for a pint was typical barter. I chuckled to myself when someone would ask to "borrow" a smoke. Really? Did they have any intention of returning all or part of it?

My son was two when we arrived, and one of the first things we bought was an English pram, a baby carriage in the states, but it had a short-sided bed, big sun and rain roof, and extra-large wheels. I would love to know how many miles I walked pushing that pram with my son riding and then him with the baby. One day, my son insisted on walking and wasn't keeping up very well. I turned around and said, "Come on Bumbles, hurry up." Two older ladies were walking nearby, and they glared at me. I had no idea why. In my mind, Bumbles was like a clumsy character in a cartoon; in theirs, it was an ass or as they called it, arse. I didn't call him that after I learned the meaning.

It didn't take long to learn the "telly" was the television and to call someone on the telephone was to "ring," "buzz," or "bell." When you overheard someone say they would "knock their friend up" in the morning, they would be by to visit or pick them up to go someplace. Mind you, I was nineteen at the time, and almost everything could be a sexual comment if you heard it out of context.

We went to the NCO Club (Non-commissioned Officers) on Lakenheath Air Force Base one Friday night and took an English "bloke" with us named Barry. In front of my husband, he asked me if I wanted to "smooch." My answer was not friendly. He was crestfallen and replied, "Geeze, I just asked you to dance." We, that could hear over the music, had a good laugh, and Barry and I danced most of the evening as my husband wouldn't.

The most memorable embarrassing moment concerning language came during our third year "across the pond." I watched a television show whose premise was to solve people's problems when they couldn't, usually having to do with something they purchased and a store wouldn't accept a return. A store named "Knockers" had ordered labeled shopping bags, which had been printed with the word "Knackers." As a naive American, I thought the name Knockers was offensive because who would call their store the slang word for a woman's breasts, but I had no idea what Knackers meant. The printer refused to reprint the bags at no cost claiming a handwriting problem on the order, but obliged after the TV show got involved. It took me a few times to the pub to remember to ask what the wrong word meant. Of course, when I thought of it, the female bartender was a few stools away, and I shouted, "Hey, what does Knackers mean?" The pub went stone quiet, and all eyes were on me. The bartender finished what she was doing, walked down to me, and whispered, "It's a man's balls." Whoops. With the look on my face after being told, the bar erupted in laughter. I joined them. What else could I do? It didn't help that I've never been a quiet person, so asking for meanings was done a bit boisterously. My husband instructed me to ask what words meant in a soft voice from that point on.

We learned bangers were sausage but were nothing like a good old hotdog, nor did they taste like American breakfast sausage. We almost counted the months until we returned stateside to have a hotdog that tasted like the ones we were used to in Western New York State. The commissary, base grocery store, carried hot dogs but not the brands we were familiar with. We also missed Kentucky Fried Chicken and other fast food flavors. Today in the UK, fast food places are open everywhere, unlike in the 1970s. We did enjoy many trips to the "chippy," which means the fish and chip shop, for fresh fried fish and French fries seasoned with malt vinegar. You could even pick different types of fish. What we know as potato chips were called crisps in England.

Ways of doing things, like fishing, differed from where we grew up in the type of fresh bait used. My husband would put our son in a backpack carrier and head down to the river that ran through the town of Thetford, Norfolk. He came home one day to tell me the older men he talked to were using maggots as bait. They would put the maggots inside their cheek to get them warm and wiggly enough to make a hook move once put in the water. I was told the men's cheeks would protrude and be in motion, and they would talk away as if nothing was in their mouths. Even visualizing that made me gag. We never tried it. I'll admit I'm a Mom that thought all kids should eat a little dirt, but I didn't want them to start picking up maggots and popping them in their mouths if they happened to see one someplace.

My children had to learn to ask for biscuits, not cookies. Meat pies looked like our turnovers and were filled with meat and veggies, and fruit pies looked the same as our dessert. Goulash was beef stew, and the Brits we knew well weren't familiar with making goulash the way I did with tomatoes, hamburger, and macaroni.

When I went to the commissary, I put the groceries in the "boot" or trunk of the car. And my son loved to see big trucks on the road, but he learned they were called lorries. If we went shopping in the town we lived in, we went to the butcher shop, the greengrocer, the stationery/book store, and the bakery. I still have and use the paperback Betty Crocker cookbook I bought in Brandon, Suffolk, England. It's 49 years old.

In the 1970s, small houses with thatched roofs could still be seen in the countryside and some smaller towns. Stone buildings could easily be

700 years old and hard to heat. Apartment buildings, or row houses, were often three stories, but the apartment was only one room wide and three stories tall. The entry and living room were on the first floor, the kitchen and bathroom on the second, and two bedrooms on the third. I'm not sure I would like all those stairs at the age I am now.

I enjoyed living in England and have many memories of specific places and events. I have friends from Thetford, Norfolk, through an international writing group, and they tell me it has changed little in the past 45 years. I would love to visit with my daughter, who was born there, and her son so he could see where his mom started life and I could see if it looks as I remember.

Tim Hansen

Tim Hansen is an Army veteran and cofounder of the Rochester Veterans Writing Group. He has worked as a public affairs officer in Afghanistan, Kuwait, and the Pentagon, and currently works as an editor for U.S. Immigration and Customs Enforcement.

The Hero Before Me

By Tim Hansen

My dad, U.S. Marine Corps, 1951-52

Once I asked him if he knew of any heroes, and he replied all the real heroes were dead. Our conversation lasted for less than a minute, and during that awkward pause, he took a sip from his vodka martini and switched the topic to the deplorable state of the economy. The war he fought had long since passed, but I learned from others about his service, his courage, and the hardships he faced.

He was born in Rapid City, South Dakota, and grew up in a family of two brothers and five sisters. His parents struggled to make ends meet, but they all learned to depend on and help each other. Christmas gifts came from the Salvation Army. Yet, he spoke affectionately of those meager years and never complained about his early years in the 30s and 40s.

In a way, poverty taught him how to survive. He learned to fight early on and to defend himself when facing hostile gangs of Native American kids over crossing into their territories on the other side of town. It didn't take much to provoke a good scrap. In the scheme of things, these skirmishes may have helped him in later becoming an athlete and qualifying as a running back in high school football.

His high school sweetheart was a quiet beauty named Pearl, the daughter of a Methodist minister. He surprised me once when he said he considered going into the seminary to become a minister and then to marry Pearl. I say that because I never saw him open a Bible on his own volition nor mention that he loved another woman before my mother.

A little later, he went off to college to study geology. At about the same time, war broke out in Korea, and the U.S. mobilized troops and equipment to defend the South and to block the communist advance from the North.

Within six months, he found himself assigned to F Company, 2d Battalion, First Marine Division as an infantryman. He spent 18 months on the front lines of Korea when he was all of 21 years of age.

He said he chose the Marines because he figured his chances of survival in the Marines were better than in the Army. I had trouble understanding that rationale since both Army and Marine units fought on the same front line, against the same enemy.

His unit fought, marched north, and managed to survive. They patrolled at night and held the perimeter. During the day, he often watched aerial dogfights between Air Force F-86 and North Korean MiG-15 jets.

What is not commonly known about this forgotten war are the Chinese infantry units that crossed the Yalu River in support of the North Korean army and engaged American units and the many tunnels the North used to move troops quickly to establish strongholds. His recollection of the horde of Chinese cavalry charging at his unit remained as vivid for him until the day he died.

One night, his platoon leader chose him to go outside the wire. Equipped with only an M1911 pistol and a flashlight, his mission was to clear out the tunnel just ahead of his platoon. He was terrified, but he did it. He came upon the enemy, but the soldier was already dead. He looked at his smooth, round face and realized they were both the same age. He searched his pockets and pulled out a photo of the man's wife and child. He never forgot that night, and it was just the beginning of many more of such tunnel clearings.

Amazingly, he survived and returned home. As soon as he stepped off the ship in San Diego, Marine recruiters offered to send him to OCS and back to Korea, but he had had enough and politely declined.

He returned home to Rapid City, worked odd jobs, and looked for opportunity. When Hollywood came to the Black Hills to film *Chief Crazy Horse* in 1954, he auditioned, passed the screen test, and played the role of a cavalry scout. He made $25 a day that summer.

Eventually, he found a good job with the U.S. Border Patrol, married a nurse from New York, and settled down. Together, they raised four kids, and he worked his way through the ranks to upper level management in the Immigration and Naturalization Service.

That man was my father.

The Horseman Passed By ...

By Tim Hansen

I still see their faces after fourteen years and wonder why they died so young. When I sat down behind my desk that August night in the tactical operations center, also known as the TOC, not too far from the flight line at Bagram Airfield, I went through the routine motions of reading media updates and answering emails from my higher-ups in Kabul, preparing for the battle update brief, and then following the various missions for a final press release by morning. The usual buzz of conversation and clicking of keyboards filled the amphitheater-like room. The giant screen on the wall in front of us came to life with infrared images from high-altitude drones over the quiet Afghan countryside and villages.

Then, around 1 a.m., the team with the mission over on the eastern side of the country radioed enemy contact with small arms fire. We all envisioned a few warning shots fired, and then followed by insurgents scampering off into the veil of night. But this time, they didn't run away. When reports of a full-blown firefight came in, all talk ceased in the TOC. I kept notes on how many we had out there and got a better fix on the team's location. Word of the first casualty hit all of us like a blow to the head. You think they are so well trained, all Ranger and Airborne qualified, that they could survive nearly anything. I needed more detail on who got hit. The operations desk couldn't be bothered and everyone kept their eyes transfixed on the giant screen.

The fighting only intensified with each passing hour. I wanted to grab an M4 and head out there, because I needed to be out there with them in the thick of the fight. I wanted to scream. Where did they all come from? Was this our Little Big Horn? Then, we received report of a second casualty. The TOC fell silent for a moment.

The group commander stood up a few minutes later, and I watched him walk slowly up the stairs to his office. I later heard he personally called the families of the men who were not coming back. The battle raged until dawn. The team missed its exfiltration time and would have to remain in place until the following night.

I didn't know these NCOs, yet I connected with them that night and I felt like they were very much a part of me. And then, I felt a deep loss as I hammered out my press release on this horrific scene, of what just happened, of the lives lost, and of the lives forever changed by this event. It was like I saw all of them before my eyes and then nothing but a gaping and dark abyss. I recoiled in my seat and looked over at my teammate, another 05, career infantry. He sat there somber and stoic, devoid of emotion. He had seen it all before somewhere in Iraq. His stony face reminded me of lines from William Butler Yeats:

By the road an ancient Cross
No marble, no conventional phrase,
On limestone quarried near the spot
By his command these words are cut:

Cast a cold eye
On life, on death.
Horseman, pass by![1]

Working through that night completely drained me and nothing numbed my pain of our loss. I turned 50 that day but realized those men who died, both in their early 30s, will never see another birthday. It was the birthday I will never forget.

---

[1] W. B. Yeats, "Under Ben Bulben," in *The Collected Poems of W. B. Yeats*, ed. Richard J. Finneran (New York: 1996), 328.

So, How Did Your Soldier End Up At Landstuhl?

By Tim Hansen

Camp Arifjan, Kuwait is an acquired taste for both soldier and civilian. By midmorning, the temperature is usually at 115 degrees Fahrenheit and the sun beats down on any living thing mercilessly. In the shade of the warehouses, workshops, and air-conditioned offices, sergeants ensure their specialists and privates drink their bottled water every twenty minutes or so. Such was life on this U.S. Army post when I was there in 2012 and so it has continued to the present day.

With the morning's battle update brief behind me, I could take a few easy breaths before starting the morning production meeting with my team of photographers and reporters. What excuses would I hear today, I wondered, to evade work. I heard the door open and shuffling footsteps. It was my team sergeant, Sergeant First Class Milton Banks.

"Good morning, Sergeant Banks," I said enthusiastically and expecting his endless litany of issues for the day.

"Good morning, sir," Banks said in a serious tone. Banks looked at life as a series of trades and deals and he excelled as a deal maker on post.

"Did everyone make it through the night safely without any heat injuries," I asked, half sarcastically.

"Well, sir, we have one soldier at the TMC for lower back pain," Banks explained. TMC is the acronym for Troop Medical Clinic. Veterans up to the Vietnam War knew it as sick call.

"Let me guess," I interjected. "Could it be our Specialist Schrodinger?"

"As a matter of fact, yes, sir," he answered. "How did you know?"

"Call it officer intuition," I said. "What seems to be the problem with our man?"

"He woke up during the night and had difficulty urinating and then felt a sharp pain in his lower back. He groaned for the rest of the night and kept Harris awake. Harris is now out of the picture because lack of sleep causes his PTSD to kick in."

"So, Harris is down as well," I mused. "Well, there goes my photo shoots of the motor pool over at the Minnesota National Guard. You know

... you and I should go over there and do the shoot. We'd probably do a better job."

"Well, there is more to report on Schrodinger," Banks said. "The PA thinks Schrodinger may have kidney stones and wants to medevac him back to Germany because Landstuhl has the resources to do more testing."

I looked directly at Banks.

"Do you think our man really has kidney stones? Last month, he had bone spurs in his left ankle and the PA wrote him a profile for no running and before that he claimed he had an allergy to sand. And when he is well enough to work, I still hold my breath. Do you remember how we sent him over to take photos of the sergeant major's retirement ceremony and he forgot to load the memory card into the camera? I mean ... can we just send him back to the Signal section?"

Banks stared back at me, taking a moment to find the words to explain Schrodinger's behavior. We both knew the problem was that no one within the command wanted Schrodinger, and so the sergeant major dumped him unceremoniously into the public affairs section.

"The TMC has already reserved a berth for him on the next flight to Germany," Banks continued. "As soon as Schrodinger heard the confirmation, he called his wife to update her on his condition. From what I understand, his wife took the news pretty hard and wants to fly to Germany to be by his side."

"For kidney stones?" I exclaimed in disbelief. "And, I suppose someone is looking into cutting permissive travel orders for Mrs. Schrodinger to fly from Fort Bragg over to Germany."

Banks looked up at me with astonishment, as though I had said something deeply profound.

"Well, as a mater of fact, sir, yes. The Family Readiness group back at Bragg has been making phone calls to get her on her way."

"Will wonders ever cease?" I asked and looked down at my outline for the next Flapjack Express news show. (Flapjack was the command's mascot, an ornery, 20-year-old mule.) "How long do you think he will be in Germany?"

"At least two weeks," Banks shot back confidently. "And if he receives treatment, it could be a month before we see him."

I thought for a moment and saw an opportunity.

"Any chance of a medical discharge?" I asked and relished the possibility of the Army without Schrodinger.

"Not likely, sir," Banks replied with a slight tone of despair. "He will somehow recover and return to the section with new ideas and a heightened enthusiasm before work loses its novelty for him and the issues begin again."

"Oh, Sergeant Banks, you know our man all too well," I said with a full measure of sarcasm.

"Just another day in this man's Army, sir," Banks countered. "And, by the way, sir ... after I visit Schrodinger tonight, I will be heading up to the press camp to supervise the media desk for the next week. With Schrodinger away, the team will be able to get your interviews and put together the next news program."

Banks had such a way of inserting his next career move amidst all of the other issues, that I could only smile and wonder how we ever manage to get anything done at all.

"Understood, Sergeant Banks, but tell me, what should I tell the chief of staff when he asks me how Schrodinger ended up in Landstuhl?"

"Well, sir, I leave that up to you."

RVWG
VERITAS PER VIRTUTEM
MMXIV

Joe Mele

Joseph Mele Jr. retired from software test engineering in August of 2018. Since then, he's had time to discover the stories in the hundreds of letters that were found when his mom passed on in 2012. The letters date back to 1934. Many of them are from the WWII years when Joe's father was the surgeon with the 749th Tank Battalion fighting through France, Belgium and Germany. It has become Joe's mission to provide background and context for the letters, and to tie them into a cohesive narrative. The Rochester Veterans Writing group provides insight into a soldier's experience, as well as friendship, support, and healing. He considers it a privilege to participate actively in the group.

November Snow

By Joe Mele

The Texas sun ascended its timeless arc and staked its scorching claim on the land. The hottest, muggiest July in decades sweated the truth from the people of Brownwood—the righteous, the sinners, and everyone in-between.

The Plymouth had proved its worth on the trip from Rochester, NY to Brownwood, TX, for Joe's November 1942 induction as a medical officer into the U.S. Army. As long as the 1941 sedan was oiled, watered, fueled, and not pressed too hard, 87 horses pushed it along with a reassuring grumble.

On July 29, 1943, the rolled-down windows brought in a breeze but no cool relief. Joe drove slowly over the nine miles of good road between Brownwood Station and the small drafty bungalow Toni called home while he was in training. Her 11:35 A.M. departure time was still thirty minutes off. He lectured her again, half-heartedly this time, on the recommended diet for her first pregnancy. Flush with the glow of her second trimester, Toni's thoughts drifted away to the sweet aromas of cookies, cannoli, and cakes in the display cases at Gruttadauria's bakery in Rochester.

The rhythms of the South had slowed their double-time New York banter. In this more measured pace, they had discussed at length the best course of action to take: Toni would return home to Rochester, NY. There, she would stay with her father-in-law Pappy, and her younger sister Lill in the big house on Jay Street, where Joe's office was. Her older brother Peter and his wife and kids lived nearby. As Peter often said: "Family helps family." The staff at St. Mary's Hospital, where Joe had completed his residency, would provide medical care.

At the station, two other travelers waited in the shade of the platform, glancing down the track for a glimpse of their ride. As the train pulled up, Toni and Joe turned face to face, each putting up a brave front for the other. Every natural impulse drew them closer and urged them to remain together. Through tears, the rational mind prevailed.

Toni was last to board the passenger car. The air brakes released with a huff, and the train slowly shrank into the distance. She looked back through her compartment window to see Joe one last time—the lone figure on the platform. The train whistled its long soulful farewell.

Lost to time, awash in a flood of memories, Joe remained transfixed, staring unfocused toward the snaking iron caravan vanishing on the horizon.

The reverie broke, and he drove, faster now, to Camp Bowie just two miles south of the Brownwood city line. New pavement rolled along the gentle, grassy hills, past mesquite and juniper, to the main gate. Beyond the sentry's hut, the Army had bulldozed the living earth into flat, precise right angles of military order. The rains had been so unrelenting throughout the first phase of construction that the crews renamed the area "Camp Gooie."

The temporary pyramidal tents had been replaced with neat rows of wooden hip-roofed barracks. Larger gable-roofed structures housed mess halls, bath houses, and recreation centers; libraries, dental clinics, and chapels; the Red Cross, a post exchange, and a hospital. Col. Frank E. Bonney commanded this city of soldiers from the highest remaining hill.

As he drove on, Joe saw the parade grounds all trampled down and the training fields off in the distance, where men and boys marched and drilled and tested each other in hand-to-hand combat; where unrelenting instructors hammered recruits into soldiers—success not guaranteed.

Back in his quarters, the wooden floorboards creaked as he walked over to his bunk and stretched out. Lying still, he swore he felt the rhythm and sway of the train relentlessly pulling Toni northward. With each passing second, a fevered thought drummed louder and louder in his mind: *"The war, the war, this goddamn war will end me and my dreams."* Hours elapsed before he got out from under this curse. "*To be home again...*" he prayed.

At 8:00 AM the next day, he penned this note to Toni:

Dearest Toni,

At the station yesterday when you left for home, I had quite a time trying to compose myself but finally, I did it. I got back about 5 minutes after 12 and didn't feel like having any dinner. I felt quite blue all afternoon, but last night

I felt better. Not many of the women were here so it was just as if you were in Brownwood.

Herbert got his captaincy last nite finally + we put our 3 bottles out + they were gone in half an hour. I had 2 drinks + stopped. The rest of the boys did quite well for themselves. Classes in International Morse are being given every night we are here starting tonite. Pretty soon we will all be going around mumbling dididida + didadidi, etc.

I hope you are having a good trip + don't cry all the way home. You don't know how much I hated to send you home but it was for the best. If I was only concerned with myself I would have kept you here. It will be harder for me because you will have all our friends and relatives at home + pretty soon Joseph Michael Jr or Joanne to keep you company.

I am making this a short letter + hope to write another later in the day. Loads and loads of love + kisses for you + Jr + the rest of the family.

Your admiring + loving husband,

Joe

****

Toni's train rolled to a slow stop at the Rochester Station, where Peter, his wife Eva, and their children awaited. Trip-weary Toni was buoyed by the smiles and chatter of nieces Renee and Marion and Nephew Anthony—so much news to tell and so many questions all at once. On the ride to Jay Street, they caught up:

Renee, in full sixteen-year-old bloom, had socked a cheerleader from John Marshall High School with a megaphone. He had made a crude comment about her body.

Marion, fourteen, saw "Arabian Nights" with John Hall and Maria Montez. "It was good," she reviewed, "but I didn't think it was right because there wasn't any kissing in it."

Anthony, the youngest, got a new clarinet. He was a natural—hardly had to practice at all.

Louie, the oldest, had written home but couldn't reveal his location—just that he was on a Navy ship and in good health.

After the short drive, Renee, Marion, and Anthony swarmed Toni like puppies as they entered the house. Overcast skies dimly lit the living room through the north-facing picture window. Toni eased herself into a brocaded wing-back chair, smiled, and basked in the comfort of familiar voices. With a few drowsy nods, she gave in to sleep and a sweet dream of days past: Curled safe in her mother's lap, four-year-old Toni consumed whisper-sung confections of Italian songs. She dozed in the chair undisturbed for an hour and awoke disoriented but refreshed. She slowly came to herself and, despite her separation from Joe, felt a peace and a certainty that "*This is where I should be.*"

Out of Joe's sight and abetted by willing family, she indulged her sweet tooth and sated her ravenous appetite. One of her letters became a confessional, recounting to Joe the entire delicious apple pie with ice cream she consumed in one day. Nothing Joe could do but scold her the next time he wrote.

****

In Texas, Joe and the 749th Tank Battalion gave up the relative comfort of Camp Bowie for maneuvers in Louisiana.

Before the last decade of the nineteenth century brought railroads and lumbermen to the state, stands of Long Leaf Pine, forests of mixed hardwoods, and Cypress swamps canopied 85% or so of Louisiana. Having depleted the forests of the North, a plague of loggers set their steel crosscut teeth to the tall, straight, stately southern pines. Teams of oxen and steam skidders piled logs high on railroad cars destined for expanding markets. Scattered clumps of inaccessible trees among vast fields of stumps and spilth gave silent witness. When nothing worth cutting remained, the wastelands were abandoned to recurring wildfires and feral hogs.

Inspired, driven conservationists rallied a movement to restore the woodlands, culminating in the establishment of the Kisatchie National Forest—hundreds of thousands of acres about halfway between Shreveport and Baton Rouge. In the Depression years of the 1930s, the Civilian Conservation Corps led the effort, building roads and bridges, constructing levees and fire towers, and planting pine seedlings. The nascent woodlands rose into an expansive demonstration of the stages of pine seedling growth: silver grass, bottle brush, and sapling.

In 1941, Camp Polk was carved from the fledgling forests of the Kisatchie. There, the Army maneuvered troops through the fields and hills, hardening the GIs for battles in Europe and the Pacific. Tanks, trucks, jeeps, and all the preparations for war weighed heavy on the land.

In the fall of 1943, the 749th Tank Battalion took its rotation at Camp Polk for a series of what they called "problems." On October 13, Joe put pen to paper and described a night in the field to his wife:

Dearest Toni,

Have already had mail call today but I didn't get anything but a newspaper. I know I will get some tomorrow. Last nite was the most miserable night I have spent in the army. We arrived at our new bivouac area about 12:30 A.M. + stretched out to sleep on the ground. About 2 or 3 A.M. it started to rain + I was too sleepy to move. I pulled the cover over my head + tried to go back to sleep. But the water kept creeping in + little puddles started to form around my back + my buttocks. I moved away + stayed dry for a while but again the puddles formed + eventually, I decided to put my coverall + raincoat on + make a run for the ambulance. When I got up to dress my coveralls were soaked, so I just put my shoes on + carried the rest of the stuff + let my bed get soaked. I ran for the ambulance + there was a full house but I managed to get in. We huddled there with the heater on for about 2 hrs. Then I managed to put on my wet coveralls + go out for some breakfast. (A cup of coffee + an egg sandwich). I am thoroughly disgusted with the medical service in the Army. I sent Lt. McNaly in day before yesterday w/ a temp of 102 + possible beginning pneumonia. They sent him back yesterday as cured, without even examining him or giving him any medication. He still feels rotten but says he will carry on till he drops. The army doesn't want medical officers, it wants men who will sign forms + who will send everyone who is not dead back to duty. That isn't giving the men the proper treatment. I feel so disgusted that I am hoping to get a little something wrong so I can get a C.D.D. I feel that I am not doing any good at all in the army after that experience. I keep thinking of home + how happy we could be. Home where I could really practice medicine doing people some good. Home where my heart is and where my happiness lies. Home where I can be clean + sleep in a bed. Home where my wife would put her arms around me and hold me tight + I would warm her cold tootsies.

There is nothing in the army that interests me or makes me feel I am using my medical knowledge. All I live + dream + think about is you + Jr + home.

Love + kisses + a great big squeeze for you.

Love Joe

After maneuvers, the 749th took the next step to overseas deployment—transfer to the largest training and transshipment facility in Texas—Camp Swift. Joe was semi-confident he could wrangle an emergency leave when the baby arrived. The army had other plans. He was denied permission to witness the birth of his first child. One day shy of Christmas, 1943, Toni brought their daughter Joanne into this world.

That Yuletide Eve, at 8:30 PM, Joe wrote this letter to Toni in Rochester:

My dearest darling sweetest wife and mother of our child,

Now at last I can say that + mean it. This afternoon, we went to the gym to play basketball + Herb stayed in the office because he had some work to do. About 15 minutes after we got to the gym there was a phone call for me. It was Herb letting me know there was a telegram for me. This was after I had received your other telegram this morning. I told him to open it + read it to me. At last, I learned that I was a father.

This is the first Christmas we have spent apart in quite a few years. Even when I was courting you, I managed to be with you on that great day. But this time we are 2000 miles apart. Still, I am happy to a certain extent because you have presented me with the best Christmas present I have ever received. The only thing that would make me happier is if I could be with you and our little girl "Joanne". Perhaps it is best that this child is a girl. When the next war comes, we won't have to worry about her being in the service far away from home.

It has been raining for 2 days + doesn't look very much like Christmas. I just returned from seeing Bing Crosby in "Holiday Inn." The first time I saw that we saw it, it brought back memories, + I could see you tonite resting against the white pillows, your black hair + eyes shining with happiness at the ~~thought~~ vision of our baby Joanne. How I wish again + again I could be there with you and her. Now I have 2 girls to take care of + I must take care of myself so they may never lack anything.

When you write please tell me all about it. I noticed that your first telegram was sent at 7<u>30</u> P.M. the 23rd + Joanne arrived at 4<u>45</u> AM the 24th so I judge that you had a short labor. At least I hope I am right in my deductions. Don't forget now write + tell me all about it. Describe our little brat to me. Does she look like most of them or is she pretty like you my sweet Antoinette. It is still so hard for me to believe that I am a father. But your description will help me to visualize her + then I may believe that I am. Also take some pictures as soon as you can + send them down.

To get off that subject for a minute I want to report my temp tonite. It is 99. So next week sometime I am going to the hospital to have an X-Ray+ a good chest exam. Although my temp goes up occasionally I don't feel as weak as I used to while I was home.

Now to get back to my favorite subject, my wife + baby. I guess you stayed on your diet pretty well to have a 7# Joanne. You are the best wife in the whole world + I hope + pray that our little Joanne may grow up to be like her mother. She will have the most wonderful teacher in the whole world, "her mother."

I hope you didn't suffer too much sweetheart. You were so vivid night before last as I lay in bed thinking of you. You suffered your pain so bravely but you wanted me near you + held my hand + then the pains were easier to bear. If I only could have been there. I couldn't have helped any physically but I would have given you moral support. Even from here, I tried to give you that.

So we spend Christmas Eve 2000 miles apart. But in spirit, we are as close as any 2 people can be. If only there were more of us in this world we wouldn't have to go through these trials and tribulations. We could be together and so happy.

Let me know how Lil + Dad + the tribe behaved + reacted. In other words, let me live through the whole thing in your letters. Be sure to take good care of yourself + later on get all the rest that you can get + need. I will <u>see</u> + it is only a <u>see</u> about the possibilities of a leave (emergency) after I get the report from the hospital.

Love + kisses for you + Joanne. Tell our little Joanne how much I love her and her mother.

Joe
P.S. Am sending this direct to the Hosp.
Love again
Joe

****

The New Year passed—a bittersweet remembrance, a hope for an end to the war. Mother and daughter were deemed healthy enough to leave the hospital on Monday, January 3rd. On the cold, snowy drive, her husband half a continent away, Toni gazed at the content, sleeping face of her bundled-up infant and puzzled at feeling so fulfilled and so incomplete at the same moment.

Settled into the warmth of their home, Toni and Joanne became the centers of each other's world.

Writing to Joe in her neat cursive, Toni described Joanne's expressions and antics, her feedings and baths, the shape of her nose, and the sound of her cries. As the days and months slipped away, the steady flow of mail followed Joe from Camp Swift to England. He saved the letters and re-read and savored them and through them, traveled home.

The Army's censorship rules prevented Joe from letting Toni know his location after the 749th left Camp Swift. By February, the header in his letters read "Somewhere in England."

On May 31, 1944, Joe sadly informed Toni that "*Since we moved into the field, I have had to destroy the letters that I was keeping because I have no safe place to put them. . .*" The loss of the treasured letters was one of the unacknowledged casualties of the war.

****

On June 29th, 1944, the noon sun shone dim and hazy through clouds and intermittent rain as the medical detachment of the 749th Tank Battalion left Southampton Bay, England, aboard LST #519. The choppy seas of the English Channel delivered them to the coast of France. On June 30, rainy still, they debarked onto the UTAH Beachhead. The landing area, secured and cleared of mines, tank traps, and barbed wire, was not without hazards.

The treads of a 30-ton M4 Tank threw mud and dug ruts into the sludge of a road just off the beach. Right behind it, another M4 bogged down, same as the first. The rest found a way around the slop, and by July

2, they had convoyed over the causeway that rose just above a flooded inland plain, past a truck sinking in the roadside swamp. Joe white-knuckled it in the passenger seat and Sgt. Machalsky drove their jeep, jumpy and tires spinning, to their bivouac area about six miles south of St. Jacques de Nehou.

The battalion parked and camouflaged their vehicles on the perimeter of a grassy field. The center had not been cleared of mines. There, the officers prepared for the impending operation to dislodge the Germans from their defensive positions in Denneville. Joe identified potential locations for first aid stations and evacuation points.

Within damp canvas walls of hastily pitched tents, the 749th sheltered from the continuing rain.

As the day's light faded, Joe strove to purge tomorrow's battles from his mind and lose himself in the solace of sleep, but the distant roll of artillery persisted through the night and imposed on him the immediacy and reality of his fate.

****

Over centuries, generations of farmers had cleared fields of stones, and tilled, and dug, and shaped Normandy to their agrarian ends. Seen from above, a mottled green patchwork of small plots sprawled to the horizon. Fruit trees, brambles, and shrubs grew atop rocky barriers that divided the irregularly shaped fields of pasture, or wheat, or flax, or rapeseed.

The Wehrmacht subverted the purpose, but not the appearance of these farms. Hedgerows were natural barriers to tank penetration and provided camouflage for machine gun nests. Narrow lanes were mined, and anti-tank guns zeroed in on bottlenecks. Interlocking zones of fire transformed these nurturing lands into killing fields. Snipers tied themselves to the upper branches of trees and favored officers as targets.

The Allied survivors of the first engagements with the enemy wised up fast. They welded steel blades to the front of their tanks to slice and bulldoze through the hedgerows. Their lighter tanks were faster, and American factories churned them out quicker than the Germans could destroy them. U.S. tank commanders avoided head-on encounters with the Panzers or Tigers. They maneuvered around to the side and aimed to disable the tracks of the superior German tanks. Once immobilized, shells could be lobbed at will at the sitting ducks.

Still, the retreating Germans exploited the hedgerows to inflict heavy damage on the attacking Allies. The U.S. Army advanced field-by-field, and Joe witnessed up close the human cost paid for each bloody yard.

Once free of hedgerows, they attacked and retreated and counter-attacked through hills, marshes, forests, open fields, and towns.

Alternately cynical and optimistic, on better days, Joe wrote to Toni:

29 November 1944
Somewhere in France

My dearest darlings Toni + Joanne,

No letters last night but I didn't expect any. However, I did get a package from Eva. It was the first Christmas package that I received. . .

Oh! Darling, I am getting so damn sick + tired of this war. It hurts me to see all this useless killing. When oh when will these damn Jerries realize that they have no chance of winning + surrender. Our Peace terms won't change + they should realize that...

There must certainly be a great deal of wickedness in this world for God to allow this war + to allow it to go on for so long.

Excuse me for blowing off like this, darling, but I want you to know how I feel about this whole thing. All my love and kisses to my two darling beautiful women.

Your ever-loving husband
Joe

****

On the home front, as Joanne grew from infant to toddler, Toni played a game with her. She posed a question in a high sweet voice: "Where's Daddy?" Joanne learned to smile and point to the framed photo of her father, handsome in his captain's uniform. Toni would clap and sing-song "There's Daddy!"

****

Though the war in Europe was officially and jubilantly over on May 8, 1945, six more months passed before Joe navigated the army red tape, screw-ups, and post-war chaos; sweated out the threatening prospect of a transfer to the Pacific, and landed safely in the U.S.

****

Snow floated earthward silently, almost motionless, in the gray morning sky. Joe's eyes locked on a single white flake and followed its slow hypnotic descent. It lay down and surrendered to the warmth of the black pavement, merging with the trickle that followed the slope of the driveway—the November snowfall's calm beauty a blessed anointing for his return. A few steps farther, a porch light shone on the green and white painted doorway. Behind it lay the vision he held in his mind for months, a lifetime it seemed—an image not torn by bombs or bullets or tarnished by the stench of rotting corpses.

Joe's hand was steady as he reached to open the back door. The tremor brought on by 194 consecutive days of combat was stilled. The door swung open to reveal a scene he had imagined over and over again. He paused at the sight of his wife and daughter. "So beautiful," he thought, "is this a mirage?" Not until he kissed Toni and held Joanne in his arms for the first time did he believe he was home at last.

The next day, Toni took the Blue Star down from the window, so grateful she had not needed to replace it with the Silver Star of the wounded or the Gold Star of the killed in action.

Toni staged their rehearsed game to break the ice between father and daughter. She asked the question: "Where's Daddy?" Joanne dutifully toddled to the table that held the photo of Joe, smiled coyly, and proudly pointed up at it. Even this slight was a delight, and Joe laughed out, "I'm Daddy!" In time, Joanne learned the Daddy photo and this stranger were the same person. Joe loved to tease Joanne about pointing to the photo instead of him. The often-repeated story became part of family lore.

They invented excuses to deter visitors. "La dolce far niente" (the sweetness of doing nothing) filled the next week, and the time passed all too quickly. In his strong-willed way, Joe tried to put the war behind him, but the soldier within was not easily exorcised.

From a fitful sleep, he would sit up with a start, waking Toni. "What is it, Joe?" Toni asked drowsily. Sometimes it was a rumbling half-track bearing down on him; or the bone-shaking percussion of exploding shells.

Reading the paper at the kitchen table, he might suddenly shudder with disbelief, expecting to find himself back at a First Aid station stitching up a wounded soldier. He might startle Toni with an angry outburst triggered by a trivial mishap like knocking over the salt shaker. Over time, in quiet moments, he unburdened to her, recounting incidents that weighed heavily on his mind. And the war fell away slowly, in bits.

He told her of American soldiers Nazis had "interrogated." Acid was injected into the buttocks of one GI, and boiling water was poured down the throat of another.

He described an incident no official action report had recorded: "The weather was turning milder. The Germans were retreating, and we were advancing fast. We were taking down a first aid station and packing up the ambulance for another move. An infantryman came around a bend in the road. He was young, but I could tell he had seen some real combat. He was marching four German soldiers ahead of him at rifle-point. He spotted an NCO and shouted, "Sarge, whaddo I do with these prisoners?" The Sergeant shouted back, "Take'm to the POW camp, private."

"Schnell, schnell," the GI barked and herded the Germans down a trail into the woods. Twenty minutes later, I saw the same private return—alone. There wasn't a POW camp within fifty miles of us."

Disbelieving her first thought, Toni asked, "he let them go?" "No," Joe replied softly, not looking at her, "He shot them."

Joe carried a snapshot in his wallet, a three-by-four wartime still life. When doubts crept in, and he questioned his own memory, he had proof. At first glance, it might be scrap lumber piled in front of a concrete block house or discarded manikins. A closer look revealed the human origins of the helter-skelter jumble. Some still wore the coarse, loose-fitting, striped pants and jackets that designated their inmate status. Others were unclothed, skin clinging tightly to bones and tendons.

He hid it from Toni until she asked if he'd seen a concentration camp. She'd heard reports. Reluctantly, he revealed the photo. After a dumbstruck pause, Toni said simply, "Throw it out." Joe didn't explain why he needed to keep it, but he refused her order. Toni relented and placed the ungodly memento upside down under a larger photo in her scrapbook of Joe's Army career.

****

Side by side, in their bedroom closet, hung Joe's white lab coat and olive drab officer's jacket, all cleaned and pressed and at attention, neither worn since his return home. Each morning, as he pulled casual pants and shirt from the closet, the lab coat silently beckoned. Three weeks passed before he gave in and donned the white coat. It felt good. The weight and smell of the cotton fabric brought him back to the pre-war life he and Toni had begun. He ambled to the parlor and unlocked the sliding door that led to the waiting room of the office suite. His footsteps echoed as he walked slowly across marbleized gray linoleum floor tiles to the examining room. The faint, familiar smell of isopropyl alcohol hung in the air. Toni had kept up an antiseptic cleaning regimen. The tray of needles, forceps, hemostats, and scalpels sat ready for sterilizing in the autoclave. A full roll of examining table paper waited to be pulled out for the next patient. After witnessing death on such a grand scale, how puny and futile these tools of the healing arts appeared, yet he looked at them more fondly now.

At the far end was the small room that held his gray metal desk. One wall displayed his framed diploma and license. The opposite wall held an illustration of a man half fully fleshed, half peeled away to expose muscles, ligaments, arteries, organs, and bones. The lone window faced east. He raised the beige metal Venetian blinds to reveal snow covering the side yard, drifting higher around the southwest corner of the black chain link fence. With the stark brightness of the scene a blank canvas for his imagination, he envisioned his daughter laughing, barefoot, arms outstretched, running wobbly in long green grass toward him. Their first Spring together would soon be upon them.

He sat down in the padded swivel chair and looked over the stationery, envelopes, and prescription pads neatly arranged on the desk, and back issues of the *Journal of the American Medical Association* stacked on a side table.

"Today," he resolved, "I start practicing medicine again." And he felt a new, more complex, even hopeful self emerging.

John Steele

Master Sergeant John Steele served in the Army and Army Reserves from 1985 until his retirement in 2015. He entered the Army as a combat engineer, worked his way up to squad leader and platoon sergeant, and eventually worked in positions as an instructor and in administration. On the civilian side, Mr. Steele taught Special Education for eleven years, and after 9/11 worked for the Department of Defense for another decade. He started dabbling with writing in 2011 during a year-long tour in Iraq, and joined the Rochester Veterans' Writing Group in 2015.

Fade Away

By John Steele

We don't want our soldiers, either young or old, to fade away. Unfortunately, I think they already have for most of the American public. For John Q. Public, many are very quick to thank military personnel for their service. And then they do their best to forget. I remember when I was going to Iraq in 2010, many people commented that they didn't know we were still in Iraq. It was forgotten. The news outlets had grown tired and moved on to more "current" issues. Somewhere around 2008 or 2009, I can't tell you the year, but a local service member died in Afghanistan, and a local television news anchor stated that she didn't know we were still there. Why is that? How can that be?

For starters, less than one percent of the adult American population currently serve. During WWII, it peaked at about 12% of the population. In 1975, nearly 20% of the adult population were veterans. Of the current American adult population, only about 7% are veterans.

For most of the recent military, it is a family affair. I saw a startling statistic. In 2019, 79 percent of Army recruits reported having a family member who served. For nearly 30 percent, it was a parent. Yet less than one percent of our people serve. The same families are providing service members over and over. But most families, and most of the American public, don't have anyone. Most families are not represented.

That is reflected in Congress. Of our 100 Senators, only 17 have any military experience. That is above the current national average, but in 1975, 81 Senators had served. In 1973, between the House and the Senate, 73% of Congress had military experience. Today, it is barely 17%. And far fewer of Congress have any children currently serving in the armed forces. Like most American families, Congress has no skin in the game.

Do you know that the last time Congress declared war was in WWII? Congress is supposed to be the only entity that can send us to war.

Another shocking fact. There are currently about 15,000 U.S. troops stationed in "Imminent Danger Pay" areas in six different countries.

2500 in Iraq

900 in Syria

4000 in Djibouti

3100 in Jordan

2000 in Turkey

2000 in Yemen

Somalia had an additional 700 up until December 2020

Military personnel can currently get Imminent Danger Pay in thirty-eight different countries. Yet no one talks about it. No one knows it. I had no idea. Does anyone care?

Do I have a solution? No, not really. But the American citizens have to become more involved somehow. One quick way to get the American public interested is the draft, but that is not the solution. From WWII to Vietnam, the draft ensured a spectrum of people and families were serving. Today's economic incentives only encourage those without economic advantages.

An option is having a mandatory service requirement for everyone upon high school completion. It doesn't have to be military service. The Peace Corps, teaching in inner-city neighborhoods, soup kitchens, Habitat for Humanity, disaster relief organizations, or some form of compulsory service requirement would help bring people together and benefit humanity. Maybe we could tie this into free state university education, technical training, or some form of student loan forgiveness program.

Women in the Military

By John Steele

I enlisted in 1985 as a combat engineer. Being that it is a combat MOS (military occupational specialty), females were banned. Now I knew that there were females in the military. I watched M*A*S*H. I knew all about Hot Lips Houlihan. I also had an older brother who was an x-ray tech in a medical unit, and his unit was about fifty percent female. I knew there were women in support roles, but I never had much interaction with them at all for the first decade-plus of my career.

From a military standpoint, about all I knew about females was based on POSH (Prevention of Sexual Harassment). We were taught to treat any female in uniform with the utmost respect if they addressed us. Otherwise, don't interact with them. Ever. Don't talk to them. Don't smile in their direction. Don't even look at them. If you even knew what direction a woman was in, you were wrong. Women were taboo. If you deviate one iota, you are jeopardizing your career.

I remember the first time I watched the 1996 movie "Courage Under Fire" with Denzel Washington and Meg Ryan. In it, a female helicopter pilot is being considered posthumously to be the first woman ever to receive the Medal of Honor. The Lieutenant Colonel leading the inquiry notices some inconsistencies in the first-hand reports of the incident and probes further. In a subsequent interview, the most gung-ho soldier of the bunch, an infantry grunt, confides that the female pilot was not, in fact, brave and heroic, but she cracked under pressure. She was a liability, and her crew was lucky to survive her failed leadership. And in my head, I was like, "See! That is what happens with women in combat. That proves that they don't belong." Yes, I know, it is Hollywood. It's fiction. Not real. And spoiler alert: In the movie, it ends up that she was the hero. Macho-boy was lying.

I was raised to be chivalrous, to always treat women gallantly. And not just women, be considerate and courteous to everyone. But especially women. At the same time, my parents instilled in me that females can do anything a man can do. And that we can do anything that a female can do. My Mom had four boys in a row. She was sure we knew how to cook and clean and do laundry. But then I spent years in combat engineer line units. I am somewhat abashed at the archaic mindset that had crept in.

I taught Special Education for a number of years. Getting kids to write was like pulling teeth sometimes. For one assignment, I had an article where there was a guy who was suing the restaurant chain Hooters so that he could be a waiter. High schoolers always thought it was silly and frivolous. So, we would list reasons for and against letting males be waiters at Hooters. Reasons for were always quite thin, but they came up with all sorts of reasons why men shouldn't serve at Hooters.

"It's always been just women waitresses there."

"Men can be waiters in other places, just not Hooters."

"Men can perform other jobs at Hooters, just not as a waiter."

"Society doesn't want to see a male waiter at Hooters."

"It's a place for women. It has worked. There is no reason to change it"

"It would be stupid for a male to want to be a waiter at Hooters."

"It's just wrong."

"Hooters is a place for men to go and look at girls, not other guys."

"A man would not get the tips that a girl would get there."

Someone invariably says while giggling, "The guy would have to wear the tight t-shirt and skimpy short shorts."

For homework, I would have them write an essay on whether they think males should be allowed to be a waiter at Hooters or not. I never had anyone argue that males should be allowed to.

The next day, I would have another article, this time discussing whether females should be allowed in the U.S. military. The classes are generally split on their opinions, but most would be in favor. Then I start playing devil's advocate. I start throwing some of the same arguments that they came up with about males working at Hooters at them for women in the Army.

"It was always just men in the military."

"Women can work other places."

"Women can perform other jobs in the military, just not in combat."

"Society doesn't want to see a female in battle."

"It's a place for men. It has worked. There is no reason to change it"

"It would be stupid for a female to want to be in the military."

The students' jaws start dropping. Then some misogynist will start beating his chest and most of the girls will start losing their minds, and a heated debate ensues. Some get furious. Some get perplexed. But they all start thinking just a little bit differently.

My students knew that at that point I was in the reserves, and they always wanted to know how I really felt. After the assignments were completed, I'd tell them that there are lots of women who are capable of performing in combat. I was sure that the women can handle it. But I wasn't sure the men could. And I didn't think society at the time was ready. This was in the mid to late 1990s. In 2003, PFC Jessica Lynch was captured in Iraq. Having a female prisoner of war was discordant with the American public and created quite a stir.

Later in my military career, I did end up in units with females, though never in a line unit. Some females were extraordinary soldiers, and some weren't, just like some males were extraordinary and others were not. Society and the American psyche have progressively been more tolerant of females being on the front lines. The chasm has certainly dwindled. Towards the end of my career, the biggest concern I had was if I were in a dangerous situation, could I take gender out of the decision-making process. My chivalrous upbringing still conflicted with the Army's gender-neutral stance.

— Rochester Veteran's Writing Group, May 1, 2021 prompt

Sick Call

By John Steele

I had an unintentional sick call event while in Iraq. My foot had been bothering me, and the medical clinic at Victory Base Complex had me taking crazy amounts of ibuprofen. My foot wasn't terrible that day, but I was running low on ibuprofen and needed to pick more up. I had a relatively slow day and thus walked over to the clinic to get more.

With my job, my mornings were hectic. I had daily urgent reports that were due later in the mornings. The weekly and monthly reports were done in the afternoons and evenings when the deadlines weren't constantly looming. So if I got a quieter moment, it was in the afternoon. It was early spring. It was warm but not excessively hot. Sunny. Breezy. Not a cloud in the sky other than the black soot wafting up from the burn pits. Yet this dusty haze hung overhead. It was a typical day in Baghdad, just like just about every other day there thus far. So I told my chief that I was going to walk over to the clinic, about a mile away, and that I'd probably be back within about an hour.

When I got to the clinic, I spoke to some Specialist and told him I was just picking up some ibuprofen. He asked how I was feeling. "Fine." "Are you sure?" "Yeah, why?" "You don't look so good, Master Sergeant. You look pale-ish." He wanted me to see one of the nurses, but I told him I just wanted to pick up the medicine and get back to work. You would think an E-8 would trump an E-4, but do you know the expression about possession being 9/10 of the law? He still had possession of the meds. He won.

So they had a nurse look at me, who said, "You look pale." "I don't get out much during daylight hours," I replied. She retorted, "Your breathing is rapid, you're wheezing, and you're sweating." She started taking my pulse. "Well, I just walked over..." No response. "It's about a mile..." Silence. "We are in the middle of a desert." When she finally spoke, she said my heart rate was pretty high. She started putting a pulse oximeter and blood pressure cuff on me as she continued with more questions. "Do you feel okay? Are you dizzy at all?" "No, I feel fine." She put the pulse oximeter on a different finger. "Are you sleepy?" "My workday has been

12-15+ hours a day every day for the past three months, and we get mortared at night. Everyone here in Iraq is tired." She left the room briefly and returned with another nurse or medical technician and a different pulse oximeter. As she read it, she frowned and told me my oxygen level was 83. "83 is passing in school."

They made me lie down, and they put me on oxygen. They're making a big deal of it, but I'm feeling fine. Are they pranking me? I worked at the medical brigade headquarters that oversaw all the medical units in Iraq. There are four Direct Reporting Units (DRUs) below me. The clinic I'm at belongs to one of the Combat Support Hospitals (CSH), which reports to me. So I deal with the Combat Support Hospital, which deals with this clinic. How would they know I was coming? Can they track me in the system to pick up meds? Who do I know at this Combat Support Hospital, and would they have one of their units pull this?

Short answer, no. Any of the other three Direct Reporting Units that I have, yes, but not this one. My brigade is a reserve unit. The three other units below me are all reserve units. But this Combat Support Hospital is an active duty unit, and they tend to have this attitude that they are a REAL army unit, and no one else knows what they're talking about. So this hospital is the least likely unit I have that would play some high jinks.

What about the other end of the spectrum? Are they just being dicks? Did I piss off anyone at their hospital? My mind was in overdrive because I didn't feel anything was wrong with me. And I had to get back to work.

The direct oxygen helped nominally, so next, they broke out a nebulizer. My pulse ox would bump to 92 or 93 immediately after the nebulizer, then fade back to 85 or 86. After the second nebulizer, they started talking about transferring me to the hospital. What really annoyed me is they weren't talking to me; they were talking about me in front of me like I wasn't there. When I heard the word hospital, I exclaimed, "What! No. I'm fine. I have to get back to work. I was only supposed to be gone an hour. It's been about two already. This has gone on way too far. I need to get back."

This time, a third nurse, the boss, came and talked to me. She was a little bit older and calmer and was wearing a pulse oximeter on her finger. She had me read it. 100. Then she put it on me. 86. She put it on one of the other medical personnel who was kinda cowering behind her. I don't think I had become that irate, but maybe my E-8 rank added a level of stress to the young lady. The head nurse explained to me that under 95 is abnormal. She had me read the pulse ox of the young lady who was wearing it. 99 She put it back on me. 88. She made a deal with me. After one more nebulizer treatment, if my level fell below 90, they would take me to Sather, the hospital on the other side of Victory Base Complex. The whole time she is making this deal, she is having different people take their pulse ox. Each one was 98 or above, and mine was in the 80s every time. I am finally convinced that they are not messing with me. But still, I feel fine.

Sure enough, I did not improve after the third nebulizer, and I am off to the hospital.

The medics at the clinic work in two shifts. There is a day shift and a night shift. Since we are mortared almost exclusively after the sun goes down, the nighttime is where all the action is. The older, more experienced medics all work at night. They get all the exciting calls, or they sit around and goof off with nothing to do at night. The day shift medics get to deal with the sick, lame, and lazy people on sick call or grumpy old Master Sergeants picking up meds. Still, they never get to deal with emergencies or take anyone by ambulance to the hospital, not that this is an emergency.

So this 18-year-old-looking medic is all excited because he gets to transport me to the hospital. I nicknamed him Jack, as in a Jack Russel Terrier. He's bouncing around chasing his tail, yapping to me how he will take good care of me and get me to the hospital real fast. I walk out to the ambulance under my own power, and I still feel fine. I meet Jack's partner, who helps me into the Humvee-converted ambulance. He looks even younger than Jack. And what can possibly be more hyper than a Jack Russel Terrier? A Pomeranian, maybe? Pomeranian boy is boing-boing-boinging around the ambulance, yapping continuously at an even higher frequency than Jack, and bounds into the driver's seat, where he has already started the lights and siren. Jack is in the back with me. He has me hooked up to a blood pressure monitor, and he is trying to get me to lie down on the gurney so he can start an IV in case it is needed at the hospital. No, I'm fine. I am sure not going to let a Jack Russel stick me.

Meanwhile, Pomeranian boy guns the engine, throws it into drive, and after a slight recoil, then a big thump, we jounce out into the parking lot. One of the puppy medics forgot to un-chock the tires. Jack is now sprawled out on the floor, swearing. I, who was once hesitant to lay back on the gurney, am now lying prone on the gurney and hanging on for dear life. Pomeranian boy meekly yips out, "Sorry," as he keeps the accelerator floored and makes a sharp left-hand turn out into the street. Jack, who was attempting to lift himself off the floor, tumbled onto the patient, me.

Victory Base Complex has a speed limit: thirty-five in the open areas and twenty around built-up areas. That, of course, doesn't apply to ambulances with full lights and sirens, even though there was little to no traffic whatsoever and no real emergency. VBC also has lots and lots of speed bumps. Or ramps. Pomeranian boy seemed to think that the further off the ground you could get the Humvee, the better. Maybe he thought we went faster in the air than with the tires on the ground. You would think it would be difficult to get an army vehicle airborne, but it wasn't. Meanwhile, I'm still convincing Jack that inserting an IV into me is not the most critical thing at this moment.

Jack spent the entire ride yelling conflicting encouragements to Pomeranian boy. "Be careful!" "Hurry up, or I'm driving next time!" "Take it easy!" "Faster!" "Whoa, watch it!" "Who taught you how to drive?" I'm thinking to myself that no one taught him how to drive; he isn't old enough yet.

For his part, Pomeranian boy did keep saying "oops" or "sorry" each time he caromed off of something or the bottom of the Humvee chassis or bumper scraped off the pavement or the next speed bump.

We did arrive at the hospital in record time. My heart was still racing. I'm sure I was even paler.

And for the first time all day, I did not feel fine.

Civilians Caught in the Crossfire

By John Steele

Since Russia invaded Ukraine, there has been a steady stream of media content from every imaginable outlet. It is so much more disturbing when that depiction involves civilians, particularly women and children.

This bombardment of information from Ukraine has been vexing, distressing, and wildering to me.

When I was in Iraq, there was an incident in Tikrit where civilians were targeted, mainly women and elders. A suicide bomber went into a mosque during services and detonated himself. People panicked and ran out of the mosque, where a second improvised explosive device, this one in a vehicle, was waiting to fragment its victims. A vehicle can hold more explosives and be that much more calamitous. There were seventy-seven casualties. But we're not done yet. At the time, there was only one functioning hospital in Tikrit, where many of the injured were taken. Shortly after the first two explosions, another massacre-er with a suicide vest entered the emergency room to prey upon more innocents. The third blast caused even more carnage.

I need to stress something. I was not there. But some of my people were. I saw the reports and pictures and talked to witnesses. And I did something I should not have done. I thought about it. Dwelled. How could someone do this? In a mosque? And a hospital? To fellow Muslims? Innocents? Multiple people plotted and implemented this scheme—then executed it with horrific results.

So what do I do now when I hear about collateral damage in Ukraine? Well, quite frankly, I try not to hear about it. It is baleful, dire. Most anything that they broadcast from Ukraine jostles me back to Tikrit. I don't need much to envision what is occurring in Ukraine. The images my mind can render far outweighs anything shown on a screen. A healthy imagination can be unhealthy.

I have spent the last decade wishing I could disremember Tikrit. Part of the horrors of war. Veterans have more urgency than most to hope for peace. People who know the consequences and fully understand the gravity of war would wish it upon no one.

Perception/Reality

By John Steele

I was asked a simple question: Is our perception our reality? I hope not.

I spent time in Iraq. Our base was filled with Iraqi civilians. We knew that some of them wanted us dead. You've heard of universal precautions for HIV? You treat all as if they are infected. We did that with people over there. We were trained to be paranoid. We were very guarded and suspicious and careful of what was said or done in front of strangers, especially Iraqi civilians, males aged eighteen to thirty-five, and any middle-eastern looking people, and all foreigners, and all vehicles, and crowds, and lack of crowds, and...

Then you come home. You have this mindset instilled in your head. And you can't turn it off. There is no switch. As a result, my perception is that Wegmans, a local grocery store, is a dangerous place. Why? The sheer number of people. The males aged eighteen to thirty-five. Anyone with middle-eastern features. Different languages. Accents. Anyone with baggy clothing. Women wearing a scarf. With all that, suddenly, Wegmans is a very dangerous place. When I walk out of there, my mindset should be, "See, nothing happened. It is not dangerous." But instead, I think, "Wow, I'm lucky nothing happened." It further reinforces my perceptions. And I am immediately scanning people in the parking lot for males aged eighteen to thirty-five and anyone who could be middle-eastern, old vehicles, weighted down vehicles, vehicles with tinted windows, and any trash lying around. All potential hazards. Paranoid? Yes. But that obsession helped keep me alive over there.

I know I'm hyper-vigilant. But turn on the television at any time. There is terrorism, hate crimes, and violence. We have politicians stirring the pot and further dividing our country. The twenty-four-hour news outlets, especially the right-wing news outlets, are all full of fear-mongering. That is not the perception or the reality I want to live in. People need to be brought together. People need to heal.

Then I see my daughter. My "Father's Hat" kicks in. My father's role often overrides my soldier's role. What is my daughter's perception? I don't want her to be afraid. I know the world can be a wonderful place. I want her to experience it, to embrace different cultures. I want that to be her perception and her reality.

Fear

By John Steele

I never felt truly afraid while I was in Iraq. I was aware, conscious, alert, hyper-vigilant, concerned, and sometimes even mad. But I had a job to do. We were programmed. When this happens, do this. When that happens, do that. And as a senior NCO, I had responsibilities to tend to. I also had subordinates and junior officers looking to me for guidance and setting the tone. We didn't really have time to think about things. It was best if you didn't think about some things.

I remember being mortared once in the middle of the day, and I was in an office with two other soldiers. There were six explosions, each successive one a second or so after the previous, and the impacts kept getting closer and closer. There was only one desk in the office, and we all had our heads under that desk with our feet splayed out in all directions. Each consecutive concussion increased in intensity. The one NCO was shaking uncontrollably, tears streaming. I remember patting her arm, holding her hand, trying to comfort her. And being mad. Pissed off. I wasn't so much afraid as angry. They were endangering someone in my company. I wasn't afraid. I was furious.

Every night outside our walls, there were calls to prayer over loudspeakers. And based on their tone, we sometimes knew we would be mortared that night. You wait for it, aware, senses on high alert, conscious of every little thing. You're ready and waiting. After the mortar attack, we all have to check in for accountability. I felt a little fear then, wondering if the people I care about are okay. First, I'd meet up with people in my section, and there was relief when I saw them. Then more concern/relief for others in the company. And it would keep broadening out. The worry wasn't for before the attack but after. It wasn't really fear.

One transport flight I was on had tracer rounds flying toward us. The aircraft I was on was returning fire as it performed evasive maneuvers. It was a total, TOTAL adrenaline rush. But there was no real time for fear. Maybe fear was outweighed by other sensations. I was so hyped up with so many other emotions erupting through me. Roused to the max, I am being shot at, yet I never felt so alive. I probably should have been terrified that time, but no.

The first time I remember being truly afraid was a week or so after I returned from Iraq. I was at Fort Lewis, Washington. One of my first nights alone, in the middle of the night, some unit was on the artillery range, miles away. The impact area was nowhere near where I was staying. You could barely hear the explosions; the vibrations were ever so slight. But that night, as I'm trying to sleep, with those vibrations reverberating through me, it was all I could hear and feel and think about. I was completely safe on United States soil. Yet, for the first time since I was deployed, I was scared. I couldn't sleep. I was hyper-alert. I had no concern. No anger. No other emotions. Nothing but fear to sweep over me and engulf me.

Photographs

By John Steele

Photos are a funny thing. I love photographs. They say a photo is worth a thousand words. But they never really catch everything.

I love mountains. There are so many great pictures of views from the top of a mountain. But, you know that if you have ever been up a mountain, the view is so much better than any picture can portray. Postcards from Iraq with a sunset, a camel, or a tank give you an idea of what it may be like there. But then, to be in Iraq with the sunsets and the camels and the tanks, it just doesn't compare. Tired soldiers sweating with grim looks on their faces. Pictures don't come close to what it looks like. Pain? A thousand photographs could never convey the pain of a seriously wounded soldier. Or the battle buddy of that wounded soldier.

Grief. A photo can capture a millisecond, a moment of despair of a deceased soldier's family member or comrade. But grief isn't a millisecond long. It endures seconds, minutes, days, and years. And just like a picture can't capture the depth, fullness, and enormity of the view from the top of a mountain, it can't capture the entirety of war. Or any aspect of war. The photo can give a taste, a hint, a rudimentary outline, and a shadow. But it is missing the story and background and experience and emotion and depth and so much more.

Watsons

By John Steele

I've struggled with an event that occurred in Tikrit in Iraq. A suicide bomber walked into a mosque during services and blew himself up. People panic, run outside, and a vehicle IED is detonated. Seventy-seven casualties. At that time, there was only one hospital in Tikrit. An hour or so later, another person with a vest walks into that emergency room and detonates himself.

Peopled planned this. If it were just one suicide bomber, I could almost understand it. Some people are whacked, lose their minds, I don't know. But for multiple people to come together to plan it out the way they did. Maximum damage. At a mosque and then a hospital. Monsters. Evil.

Before we ever went to Iraq, the army prepared us with all the propaganda, not to trust Iraqis, loose lips, and all that. So I was already wary. But Tikrit shook me. I was leery of, hesitant towards, and distrusting of anyone who could be middle-eastern. And it's lasted. I'm not proud of it. I'm embarrassed to feel the way that I do. But what kind of person could do such a thing?

My daughter and I read together. She's going into eighth grade. We just finished a Newberry Award-winning book, "The Watsons Go to Birmingham–1963" by Christopher Paul Curtis. It's a story about an African-American family from Michigan. Throughout the story, you grow attached to the Watsons, and you know how much they care for each other. They travel to Birmingham, Alabama, to see family. With the unrest from the civil rights movement, a church is bombed, killing four young girls who are attending Sunday School. In the book, the young Watson girl turns out to be okay, but I know it is based on an actual event. It took me back to Tikrit. What kind of person could blow up a place of worship? Who could target, injure and kill innocent children? What do they look like?

And it hit me. Some of the people who could do such a thing look just like me.

Unsung Heroes

By John Steele

I was asked about unsung heroes I served with in my military career. I was fortunate enough to serve with so many. I was in uniform for over twenty-nine years, in several active and reserve units, with a few different MOSs and numerous positions. Pretty much everywhere I went, I was surrounded by unsung heroes, though some were a little more unsung than others.

Two groups jump out for me from the get-go. First, I spent some time in a Warrior Transition Unit, teeming with soldiers who were injured or became ill, most of them from Iraq and Afghanistan. I made friends with one double amputee. He lost one leg just below the knee and the other halfway up the femur, and he learned to walk with prosthetic legs. Others lost limbs or eyesight or suffered burns, sexual assault, traumatic brain injuries, breathing issues, or PTSD. It was incredibly humbling.

The other group was the soldiers of headquarters, 804th Medical Brigade, whom I spent eleven months with in Iraq. We weren't kicking down doors, weapons blazing, capturing insurgents and caches of weapons. We spent most of our time sitting behind desks, ensuring all medical operations for all U.S. service members and their allies throughout Iraq ran smoothly. And they did. Everyday soldiers doing their jobs with very few accolades.

My time in Iraq and the Warrior Transition Unit was the most noteworthy, but my entire career was strewn with unsung heroes.

I first joined the army as a combat engineer. Most combat engineers took a lot of pride in their MOS, but most MOSs could be pretty boastful. However, unless a soldier is in a dress uniform, which is rare, you don't truly know their MOS. Except for infantry. Infantry sure could crank up their level of arrogance.

Most soldiers in their duty uniform look just like all the other soldiers. Awards and ribbons aren't worn on the duty uniform. Other than rank, there are only a few ways to differentiate various soldiers. One is a combat patch. Every soldier is versed and ingrained as to who has been in combat. But while soldiers are innately aware of combat patches, are civilians? Another distinguishing feature is anyone who has earned the special forces, ranger, or sapper tab. They are revered a little more by soldiers. I do not know if civilians notice or understand.

One unique feature on every soldier's uniform is their unit patch. Some active duty units get a little more than their share of attention. I am referring to airborne units. Some personnel in airborne units can't help themselves, and they point out their unit patch much more than necessary. They do their best to eliminate the term "unsung" from their hero status.

Oh! And drill sergeants with their stupid hats. I used to have respect for drills. I hated them, but I respected them. That was until I became an NCO myself and met some. They sure don't think their shit stinks. It does. For a good time, go to an NCOES school that has some drills in the class. Watch them pout when told they can't wear their stupid hats and then watch them flounder in the class because they only know how to push troops and the basic tasks learned in basic training. They know everything a private should know inside and out but nothing about the tasks necessary for Corporals and Sergeants and above.

I've pointed out an array of differences in uniform, but they do not tell the story of a particular soldier. And how often does John Q. Public see military members in uniform? I am guessing that, within the community, military personnel are seen without anyone realizing it. There may be some active duty in certain sections of the country, but reservists and national guard members are sprinkled everywhere throughout the population, as are veterans. I would argue that reservists and national guard members have it harder than soldiers on active duty.

On active duty, your entire day is dedicated to physical training and proficiency in your military career. Reservists and national guard members have to fit it all into their regular lives, along with their regular jobs, family, school, etc. Think about it. They have to be skilled in two different occupations, one military and one civilian, as well as fitting into two different societies. And one society has no clue about how the other one functions. Anyone who believes it is just one weekend a month and two weeks during the summer is mistaken. With all the sacrifices they make, part-time soldiers are definitely unsung. They may march around once a year in a parade, but they don't get the fanfare and hullabaloo they deserve. They're anonymous, living a hero's life in plain sight and no one notices or pays attention.

There is another group that gets even less of the recognition that they deserve: the military families. Soldiers signed up for this lifestyle. Spouses did not. And the children definitely did not. Active-duty families have a great deal of assistance and services available at their respective bases. Reservists and national guard members do not and are left dangling in the wind. The limited systems they have are filled with foreign processes and procedures, are inconvenient to access, and are brimming with an acronym-laced language they don't understand. Neither the military establishment nor civilian society understands the struggles and difficulties encountered by military families.

I also want to touch on some of the different eras. World War II, the greatest generation, that is quite the distinction. That is still not enough. They should be honored and revered. Unfortunately, our country's honor and appreciation for the military have waned. Korean veterans never got their props. It wasn't even called a war. Then came Vietnam, and instead of simply not getting their deserved recognition for their service, they were scorned and treated with hostility. Every time I get thanked for my service, I am saddened and embarrassed that Vietnam vets never got their due.

For all of my military career, it has been an all-volunteer force full of everyday people who have stepped forward and offered themselves to defend their country. In my book, each who has served honorably deserves “unsung hero” status.

I Know You

By John Steele

I know you.
I've never seen you before, but I know you.
I've never talked to you, but I know you.
I can't tell you what color your eyes are, but I know you.
You're much younger than I am, but I know you.
We wear the same uniform.
I may have passed by you at some point.
I hope I had a kind word for you.
But I have no idea if I've ever seen you. Still, I know you.

You have visited me so many nights. Almost every night.
If I saw a picture of you prior to the time I saw you, I doubt I'd recognize you.
I don't know your face.
I don't know your name.
I don't know your unit.
I don't know what state you're from.
I don't know your background.
I don't know your family.
I don't even know your name, though I read your name tape.
But I know you.

So many details are foggy, or a blur. Or just blank. Gone.
Your name. Your unit patch. Your rank.
Yet so many other details are so vivid.
They will never go away.
The sandbags. The concrete.
The shadows inside the bunker.
The red. So much red.
The texture. The pieces. Globs. A couple of fragments.
You looked so young. So innocent.
Your whole life was in front of you.
How could you do that?

What could have been so bad?

You come to me so often.
You never say anything.
No answers.
No reasons.

It's too late now, but I see you.
I know your pain.
Somehow I know your pain.
Sometimes I feel it too.
The desperation. The loneliness.
The hurt. The confusion.
The helplessness.
There had to be another solution.

You appear when I'm alone.
I don't have any answers.
I sometimes wish you did.
I know you.
I'm hurting too.

Cindy Blair

AG1 Cynthia (Cindy) Blair grew up with a love of God and Country as the daughter of a Korean War Veteran in Upstate NY. She felt the call to duty after the fall of the Berlin Wall and in September 1989 signed up for the Navy on a delayed entry program and worked in the recruiting office. May 1990 - left for basic training in Orlando, FL; Jan. 1991 - Aug. 1995 - weather observer at Belle Chase, LA

Aug. 1995 - Aug. 1998 - weather observer and assistant forecaster at Sigonella, Sicily, Italy; Aug. 1998 - Sep. 2002 - Meteorological Research Assistant at Fleet Numerical Meteorological and Oceanographic Center in Monterey, California. She continues to serve in her community in various volunteer and peer support roles.

For God and Country

By Cindy Blair

I often get asked why I joined the service and chose the Navy. First, my father served during the Korean War, and I wanted to follow his patriotic lead. Secondly, I grew up as an active girl scout from ages 6 to 17. Our pledge included the words, "I will do my best for God and Country." I always felt it would be an honor to serve in the military, but I hesitated because I am female.

When I was 24 years old, two things happened that changed my perspective. First, I observed a woman in her mid-70s working with me as a cashier at the Winn Dixie. I asked her, "When are you going to retire?" She replied, "I don't have enough money to retire." Her answer prompted me to think about my future. "What else can I do?" I thought. While looking for another job, a Marine recruiter stopped me on the street and told me a lot about military service. I had heard horror stories about being in the Marines, so I told him I would think about it.

Another week or two went by, and I still hadn't found any good job leads. I considered moving back in with my parents but quickly dismissed it because I was enjoying my independence, meager as it was. I made plans to visit them anyhow while I continued to look for a career path. As I approached the door to the travel agency to pick up my tickets, I noticed the recruiting offices were right next door. I didn't go in, but I thought about it some more.

After I returned from visiting my parents, I heard about the Berlin Wall coming down. I began to hear the call to service on TV and amongst my peers in the community. I approached the Air Force recruiter because I believed the rumor that it was easy duty. The Air Force recruiter said they would put me through basic training and then choose a job for me based on my abilities. Again, I said I would think about it, but I made an appointment for the next week. When I showed up for my appointment, the Air Force recruiter wasn't there. The Navy recruiter was just across the hall, and his door was open. After talking with the Navy recruiter, I was ready to sign up. The Navy offered me my career choice and gave me advanced rank due to my associate's degree. May 1990 marked the beginning of a fulfilling career.

My Most Beautiful Memory

By Cindy Blair

Growing up in New York State, I had never traveled very far west. So, I was thrilled when I received orders in the Navy to Monterey, CA. My residence was located on a hill where just a few steps out my back door, I could see Monterey Bay. The view was a breathtaking postcard picture; after the morning fog cleared up, of course. Living and working in Monterey was a pleasure I never dreamed I would get to experience. On my days off, I would go for long walks or drives along the coast. Many a scene of natural beauty caught my attention, but my favorite is one that became my most beautiful memory.

One day I arrived at Pebble Beach after a late afternoon storm had moved offshore. A larger-than-life sunset infused the sky with lavender, violet, and amethyst purples. The water sparkled as bits of sunlight danced like fairies across its rippled surface. I pulled my gaze away from the sunset to gather some of the smooth stones from the edge of the water. Their bright colors were enhanced by their wetness. I tried not to take too many, but it was called Pebble Beach, after all. I forced myself to stop at about a dozen stones with each color of the rainbow represented, plus a shiny black one and a white granite one.

When the sunset grabbed my attention again, the scene was ethereal, and it inspired me to write the following poem:

Storm Port

The heat of the sun has gone with the passing of the storm.
Clouds crumble to bits in the tempest's wake,
letting peach-hued sky reveal itself in quiet splendor.
I try to capture the moment with my pen to tablet,
but reasonable speech escapes me.

The incessant calling of the seagulls invades
the silent motion of my pen, and it falters.
I also take pause.

Fresh wind fills my nostrils with rain-soaked earth.
The rock beneath my thighs cools to where
it is almost unbearable, but still, I stay.
Entranced by the vision before me, I'm asking the question,
is the sunset woman or man?

Waves lapping at the shore capture my attention and
my pen again picks up their rhythm.
The dark purple horizon is fading into black.
I should go now before eyesight is no more.
Before I go, I ask the shore,
"Will you be the port in my storm?"

Every time I visit California in my mind, I drift to memories of Pebble Beach. When people tell me to envision my happy place while meditating, I conjure up the glorious sunsets and peaceful natural displays ingrained in my memory. Sometimes, I look at my collection of stones from that day or I re-read the poem it inspired and I get transported back to that moment for a little bit in time. Maybe someday I can return in person to revisit my most beautiful memory.

Go Kart

By Cindy Blair

A square-wheeled go-kart
can't go very fast
but a push from a friend
will move it forward, at last.

The more friends who push it,
the more times it goes 'round,
will shape it and mold it
'till its strength can be found.

He had thought he was broken
stuck on a road out of town
when new friends and old ones
showed him a strength of his own.

Now, that square-wheeled go-kart
can race along with the best
as the square wheels are rounded
and only stop when they rest.

A Veteran's View 20 Years After Service

By Cindy Blair

A war of enemies
and an enemy of wars.

A friendship of comrades
and a comradeship of friends.

A camaraderie of forces
and a force of camaraderies.

A bravado of pictures
and a picture of bravados.

A tangle of emotions
and an emotion of tangles.

A concert of heroes
and a hero of concerts.

A fellowship of coworkers
and a coworking of fellows.

A companionship of acquaintances
and an acquaintance of companions.

Lee Nelson

Lee was raised until the age of eleven in Rochester NY. His mother remarried in 1984, and this led him to Waterloo NY, nestled in the Finger Lakes region, and the proud birthplace of Memorial Day. He joined the Air Force in June of 1992 and completed one enlistment of four years. At the time he believed he required discipline and direction, and he got both. The military changed his life. He returned to upstate NY in 1998 and went to nursing school on his Montgomery GI Bill. Prior to nursing school, he was a surgical technologist and has now been a nurse since 2005. Altogether, military service included, he's been in the medical field for thirty years. He's presently an operating room nurse, and he and his loving wife Nicole have a deli, Lakeside Deli and More, in Irondequoit NY where they've lived happily and peacefully for fourteen years. In their blended family, they have three adult children.

In his spare time, Lee has always aspired to be a writer. He's presently putting together his first book of poetry and has also written several short stories, an incomplete novel, and one screenplay in desperate need of revision. His other hobbies include weight lifting, camping, landscaping, craft beer and The Bills Mafia.

Lee had to leave the military to love the military. This truth may reflect in his accounts. He's glad he did it, glad it's over, and honored to be among those who chose to serve. He thanks you for your time, and he never would've missed the opportunity to share.

Thank You for Your Service

By Lee Nelson

With ancient design
the military shouted
and berated
and scorned
and dismantled
and instilled into me
all that I've managed
to become
and it could care less
for right or wrong.
It just does what it did
and will do again
for any "serviceable"
semblance of human
it can find.

What the military does
heinously
it does best
and that goes for everything
even war never touches
as our country
never ceases and desists
war.

I'm thanked for my
"service"
in the obligatory
and thoughtful fashion
and to me
my debt is till my death
a mortal truth even I
don't entirely understand.
I don't feel the military

in my trauma
or any of my
heinous starts
in the darkest nights.
I'm thankful for all
I do not understand
but I do believe
I speak for many of us
when I tell you

me and the machine
got what we needed.
I walked when I realized
there's no mentors left
when the trail arrived
to everyone like me
looking for what's next
other than ownership
or death.

I grabbed the Greyhound
with all they'd promised
everything I'd earned
and made a life
only discernable
with the flag
in my rearview mirror
and here I am.

As you bury
the patriots you raise
your rapturous everything
that defines you
the cored heart of your
most purposed joys
and terminated
immortalities

thank me
for that..

MEPS

By Lee Nelson

People often ask a veteran about their basic training experience, and you could ask and I could happily tell you my story. Anyone who's ever been has plenty to share. It's an experience you never forget the rest of your life for a lot of painful, growth-worthy, and even humorous reasons. But you never hear a MEPS story, and every veteran shares this experience. It's not an awful or particularly noteworthy experience, but it's not a benign one either. So, what the hell, how about a MEPS story? I may just have a good one.

It all began and it all begins with MEPS: Military Entrance Processing Stations. These stations are everywhere. Depending on how rural you may live, you'll find one in the closest "big city," and if you're relatively metropolitan, you'll find one or more right at home. You'll of course never know MEPS without a recruiter, so you'll need one of those first, but he's very easy to find. When you're at the young, spry age to entertain enlistment, try missing him, I dare you. He'll come to your high school looking for you if you don't choose to pursue him. He may even be in your bedroom. He's a rather aggressive fellow. He's always looking for soldiers because that's his job. So, find him or let him find you and you'll surely find your local MEPS, and there the experience begins but hardly begins.

To my experience, MEPS in itself is a reveal to military training: get in line and shut the fuck up. We're in charge. Wanna be here? Good! If not, go home to Momma or we'll send you home to Momma when we fail you. By the way, "failure" is not a failure to take to heart. It's a failure of physical means upon their assessment, and their "assessment" is as arbitrary as failing the next Deion Sanders to conceivably fire a rifle or sit in a cubicle, or Bill Gates to run a mile and sort ordinance, let alone the uniform expectation that everyone has to be fit enough to die for their country. They come to the table with predetermined "minimal standards." There are no failures in MEPS, only impasses. If you fail, it's life telling you to look elsewhere and nothing more. I've been slowly going deaf in my left ear my entire life, and I "passed." I've never been a fate kind of guy, but I guess it was meant to happen. In the beginning, all the military is looking for is some kind of sound and physically able monkey, and they want as many able-bodied, "mentally sound" monkeys as they can find. Speak when spoken to, be the best monkey you can, and see what happens.

I remember my recruiter, MSGT Frank Koskowski, insisted I take a bus to Syracuse the night before my MEPS experience to get a good night's sleep and arrive early, and I mean early. We were there at 6am, and the MEPS world was waiting for all of us. That's another thing you learn very quickly in the military: you get more "done" by noon than most people do by 5pm. You will never sleep in when serving your country. You will not. Everything begins at an obscene hour of morning.

Now that I think about it, my introduction to military life may have just begun at the welcome desk of the Syracuse Sheraton. Yes, the Sheraton. Our taxpaying dollars put me up for the night at the Sheraton before MEPS in the morning. I suppose it could've been the Red Roof or a Travel Lodge, but the cash went to the Sheraton. It was clean and pleasant. They all are for the money, but wow was the receptionist rude! I'll never forget her. There were about six of us adolescent boys and girls chomping at the bit to serve our country, and she stood us to the side of a few other paying customers (even though we were there first) and made us wait as they were congenially provided for. As every one of these customers approached the desk and sheepishly acknowledged our presence, this rude woman assured them "they are just military and they can wait." So, we waited. We were kids. Nobody had the balls to question her.

When we were all finally addressed, she handed us all keys and said, "No shenanigans, lights out at eleven. Wake up call is 4:30. The bus arrives at 5:30. The government pays for your bed, nothing more." You might think she'd have said "any questions" after that, but she did not. She dismissed us like dirty peasants. I wonder if she didn't think she was somehow in her own way serving her country. Maybe she was. I have to admit she was spot on.

At 6 a.m. the next morning, everyone was exactly like her. Most were even worse. I particularly remember a guy I discerned was Navy (must've been the bell bottoms), and what an asshole he was. He was pot bellied, bald and bespectacled and clearly miserable with the world of his work, and we were his work. He did a lot of shouting and pointing and none of it made any immediate sense to figure it all out, but we managed the circumstances like deer on the interstate. We all somehow did as told, and after the sexes were segregated, us boys (lots of us) found ourselves in our tight whites in a cold gym the size of Nebraska. We were then assessed by some kind of practitioner in a lab coat. He grabbed faces and looked in eyes. He demanded open mouths and tongue turns to the left and right. He examined our hands and feet and everything else from craniums to spines to asses. Then some page arrived to take all our blood pressures. Yeah, shit you not. Must've been the stupidest sequence of events ever decided, or a deliberate mechanism to weed out the meek.

After that there were physical challenges, hardly difficult ones, but quite awkward requests. Who the hell ever walks on their knees across a room? That one was fun. I don't know how anyone "passes" that one, but I can't tell you anyone failed. Then there were "bear crawls." I'd done a lot of bear crawling in football practice, so I guess I had it mastered, but I'd never done it near naked. For the record, nobody needs to bear crawl in their underwear. In fact, nobody ever needs to bear crawl. I've never had a situation in life where bear crawling was necessary other than MEPS and football practice. Perhaps if a house was on fire, or to fulfill a Festivus feat of strength. I've had even less life-long necessity for algebra, but I digress. Then toddler crawls on hands and knees. Then we just walked. We walked back and forth across the gym. Then we touched our toes and bent backwards and did physical things that may have made some form of sense, and then, very thankfully, we were told to get dressed. After that, things got a little more intimate, and they got even more intimate when they told us to get naked again. By the way, I'd expected at least one or two pushups and sit-ups, but there was none of that. Why was there none of that? There was damn sure plenty of that in basic training.

Once we all "passed" the weird physical challenges, eyes and ears were next. I had expected the hearing tests to be my Achilles heel. If you've had poor hearing your whole life, you take solace in knowing your eyes are worth a shit. On came the eye exams, and everything was fine until the "depth perception exam." I had no idea what the exam even was or what they were measuring, but there it was. They had us all place our faces to the customary machine with the comfy, rubber crested binoculars on the top of it and view the endless series of prisms within the viewing space. Well, they called them prisms. Actually, the bald asshole Navy dude with the glasses called them prisms. Actually, he yelled it, and we slapped our faces to the machines. He had to repeat to me what I was looking for within these prisms, and he really didn't care to repeat himself. What he wanted was discernment. There were prisms and prisms and prisms. There were nothing but them, and each prism had four black dots. One dot within the four of each prism was at the forefront of the prism, and the remaining three were in the rear. Every one of the many prisms had a title such as A1 or B11, etc. Man, were there a lot of god damn prisms, and when prompted, you were asked which black dot in C12 was at the front of the prism, then E13 and F8 and then D16, etc. They all looked like clones to me. I didn't know one prism from another, and the pressure didn't help, but the man wanted answers right now. I can't tell you to this day if I failed miserably because my depth perception blows, or if Mister Miserable Navy Man got my goat. Well, he definitely got my goat because I still want to punch him to this day, but maybe my depth perception actually sucks. For all the consequences that exam had to my ensuing four years, I've never bothered to know better, but that exam was big for me. My "failure" proved to eliminate a lot of job options, but it's funny the job it didn't disqualify me from performing. Cleaning teeth? Seriously? I'd think depth perception matters in a human mouth, but to that man's miserable credit, I don't recall ever hurting anybody. Touché, douche bag. Maybe you knew something (probably not), but screw you anyway.

Then there was the foreboding and inescapable hearing exam. The eyes had a few exams and the ears only one, but I guess for as important as hearing goes, one will do. I did well enough in the beginning. I don't know about you, but for all the hearing I've failed to have in my lifetime, when you offer absolute silence and then subtle tones, I'm going to hear them. Even with my chronic tinnitus, I've always heard the offered exam tones, no matter what pitch. It's when the varying background noises are presented where things get hairy for me, and they sure did. They quickly made their way to conditions where I couldn't hear shit, and it felt like forever that they waited for me to respond to the tones. I can't tell you what was really happening among the evaluators that day, but I feel to this day, for the lack of a better explanation, that somebody may have understood me, or at least understood my condition. Maybe I had enough A game to make the minions push the envelope? I don't know, but I can tell you what happened next.

It ended up being a longer day for me than most. Some guy with a few stripes was pulled from his desk and tasked with taking me downtown somewhere to have my ears cleaned by a physician that spoke very broken English, but was probably happy to get paid, and must have had the time to see me. He was all business just like the rest of them, and he came after me with a water pick. Aren't those for teeth, I thought? Not today. He said little and worked in physical instruction, and put the pick in my hand, made me hold it to my ear, and on came the hydrant hose. Both my ears were given a damn intense washing, and I think my brain was pressure washed as well. I was sick and dizzy when it was all over, and in all my dizziness, the man presented me a kidney basin filled with horrid shit. He said "blek." That word goes for all languages, and the few striped guy drove me back to MEPS. He was worried I'd puke the whole drive. I didn't. I don't know how, but I didn't.

Then we returned to the ear exam. They took it from the top and did every step all over again. For the record, I'm really glad the Navy guy wasn't there, because in his absence some people seemed to care. No rock was left unturned, and when the random distractions were added, I still struggled. Somehow, I made it through and just barely. I know that because a guy's voice finally came through the headphones like the uninhibited decree of Moses and said, "Okay, you passed. Barely, but you made it." And down the hallway they dismissed me.

Then I got naked again with strangers. It was late in the day, so they were new and different strangers, but we were all equally uncomfortable. No page arrived to take our blood pressures. I guess that test was one and done.

We were all walked single file to a doorway. It felt like something out of Pink Floyd's "The Wall." What happens beyond the door? The instructors had zero interest in throwing a bone. I gathered they all came from the same school as Navy Man. With every one of us that arrived at the door, an instructor would point to the room in case anyone was a slow study. Naked and cold as shrimp, I imagined us all getting swallowed whole by an unworldly beast. That didn't happen, but what did happen everyone involved could've lived without.

I assume what happened beyond that door doesn't happen now? It can't possibly happen now. That would hardly make sense. It made jack shit sense even then. Let's just say it was an uncomfortable and cursory experience. For what was asked of all of us, I'm pretty sure the practitioner was even less interested than us. He told us what to do and we did it. By the time we were all 99.8% finished with the request, he said "fine," and on we all went to find our clothes. This was 1990. Bush Sr. was still in office. We were years away from Bill Clinton introducing "Don't Ask Don't Tell" between Jennifer Flowers and Monica Lewinsky. Where the rubber met the road in 1990, I can tell you that nobody was interested in finding a "culprit." That was at least my experience, and I only hope that it was everyone's experience. To think of it now leaves me believing that we all make the right decisions (more or less) without agendas. Are you "able bodied?" Do you want to serve your country? That's all that should ever matter if you ask me, now and forever.

By the way, we were also never asked to turn our heads and cough. Hernias are kind of a big deal. They sure can be. What a damn fickle experience it all was. There were no mental exams either, other than a questionnaire that asked us all if we've ever wanted to kill somebody or ourselves, or whether or not we've ever smoked joints or any other construable impure thoughts in the name of lord and savior, and I gather we all knew enough to lie our ways through the fifty questions with our number two pencils. I must say the killing questions still stick with me. Was there a wrong way to answer the questions? I erred on the side of caution as anyone would, but then I got to basic training and it was all about killing in the name of "freedom." It was also about properly folded socks and dust-free floors and hospital corners, but to take orders without question and kill on command were the lessons to take home. I've never knowingly killed anyone to this day, but I do love a clean house. I may love it too much. My kids certainly hated me for it. They still talk about it as they revel in their filth, miles away where my authority is null and void. I've always been a neat freak. The military just taught me how to best sharpen the blade. Whatever, my kids are better for it no matter what they'll tell you.

At the end of it all, my mother was waiting in her Buick Century. She'd been waiting a while, quite possibly hours. I can't recall, but there was no better way to play it. I had it pretty good. She could've told me to take the bus home. I was adolescent and annoyed. She probably took a lot of shit from me on the drive home. She always did. I did the same for my kids, so again don't let them tell you differently.

She didn't ask much about the experience. My mother always waited for exposition. Every smart parent learns that lesson. Don't ask for the drama. It'll come. It's always chomping at the bit. Only this time it wasn't. I didn't want to talk about it. I didn't like the Navy man or the other douche bags, and I sure didn't care for what revealed beyond the door, so none of it was anything to chat about. She picked up the little I laid down.

"The recruiter called twice," she said.

"Yeah," I replied.

"He wants you to call."

"Okay."

"I don't like him."

"I know."

"He's pushy."

"I know."

"You still want to do this?"

"I think so."

"You know we're at war, right?"

"Yes."

"What do you think of that?"

"I don't know."

"We're at war. You should know."

"Okay."

"Not okay. We're at war, and he's pushy. I don't like it."

"Okay."

"Not okay. Think about it."

"Okay."

"No. Not okay."

It went on like this all the way home. No revelations were made and I joined the Air Force. My mother learned to accept it and even grew to appreciate it, but she never liked my recruiter. Not for a second. Expect your recruiter to behave the same, and expect your mother not to like him. It's likely and obligatory. Recruiters do their jobs. You don't have to say yes.

At home the next morning, she asked about the hearing exam. She'd obviously wanted to ask the whole ride home but refrained. She was hopeful that I'd failed. She wouldn't let Operation Desert Storm die. She also didn't like Stormin Norman Schwarzkopf. She said he looked like a butcher. I continued to remind her that I was joining the Air Force and not the Army. It was all the same to her as it would be to any mother, and until the day I raised my hand and pledged my allegiance to the Air Force and The Uniform Code of Military Justice, she swallowed deep inside her and supported me lovingly but begrudgingly. She never gave me her blessing, but she accepted my decision.

Six months later, we returned to MEPS. Once again, the Sheraton was my hotel of stay on our taxpaying dime. My mother wasn't going to miss my swearing in let alone my final moments with her before I was bussed off to the airport en route to Lackland Air Force Base, Texas. She got her own room at the Sheraton on her own dime. My mother always loved a good cocktail, and she had a few, and grabbed me a six pack for my sendoff.

It was a first for the both of us, mother and son tipping a few back before the curtain called. I'll never forget it. That woman sure loved me. She needed her cocktails more than anyone in the world that night.

It Does As It's Told

By Lee Nelson

Some assignments just grab you and stoke the hell out of you. Others, the very mention of them, deflate you to the very core of your soul. Without the reveals of personal experience, there's no middle ground at all. By word of mouth, your arbitrary destination is either famous or infamous. Who doesn't want to be sent to Hickam Air Force Base, Hawaii? Yeah, exactly. You could be the most miserable soldier on the planet. Someone says Hawaii to you and you're at a complete loss for downsides as you quickly imagine yourself R and R'ing on the shore of Waikiki Beach, ass in the sand, Corona in your hand, bikinis as far as the eye can see, your young loins directing and dictating your very purpose and existence. As my grandfather used to say, "Happy as a pig in shit." Sun, fun, beers and bikinis. No down sides at all.

A close second can be said for the beers and breasts of Germany, a very different assignment altogether, but again, what are the downsides? A long enough stretch at Ramstein could allow for not only an enriching life experience in Germany, but perhaps an opportunity to see most of Europe if you play your cards right.

In the same vein, how about Lakenheath, Great Britain? Here we're talking about another enriching and foreign experience with pubs full of dartboards and bitter ales, and you can even speak the language, not to mention also play your odds at seeing most of Europe. Oh, and ya know, they have women too.

Eielson Air Force Base, Alaska on the other hand? Twenty six miles outside of Fairbanks? Perpetual daylight and perpetual darkness, miles and miles from every exceedingly rare, parka covered breast, and absolutely nowhere to go and no way to get there without a snowmachine? Let me just say that Alaska is beautiful (to say the least), but you really have to be built for it, and you really have to want to go there. Otherwise, you're going to end up bloating your liver on shots and Schlitz and mess hall grits. "See the world" they said. You can see a lot of world from Eielson, so much world you might think you're the only man in it. You gotta want that one. You sure do. Most of us do not.

How about Dover Air Force Base, Delaware? If you're not from Delaware, does Delaware sound appealing? It's on the Atlantic coast, I'll give it that, but our latest president aside, ever heard of it? Do you know anyone from Delaware? I don't know about you, but Delaware sounds like machinery to me, and while they're probably hardly synonymous, Dover sounds like Detroit. I imagine abandoned manufacturing plants with broken glass behind chain link fences flanked with weeds guarding rusty machines of a yesteryear purpose waiting to die with all the earnestness their former masters wait for another job. Yeah, I could be terribly wrong about Delaware. I don't know a thing more about Delaware than I did twenty nine years ago, but I didn't want to go there, and neither did anyone else I knew. I also immediately didn't want to go to another destination, a destination I'd never heard of in a state I knew absolutely nothing about, and a place I was going with all the declared authority of a father that says "we're moving."

I was slowly creeping my way to the tech school finish line at Sheppard AFB, Texas when SSGT Cumbi step-stopped us in formation, ordered us at ease, and announced our pending reassignments. We'd all been through a number of rounds of these reassignment announcements by that personally fateful day, and I was among a select few that felt like neglected stepchildren still salivating for the better or worse news. It finally came. I was third to last, but there it was. SSGT Cumbi dropped the previous orders in his satchel (because the airman wasn't present at the time), looked to mine and couldn't control a smirk.

"Airman Basic Nelson," he questioned.

"Airman Basic Nelson here, sergeant," I responded.

"Airman Nelson, you're going to Minot Air Force Base, North Dakota," he answered with an ominous smile. And mocking laughter ensued.

He then came to me and handed me the orders. He looked at me with sincere regard and said, "Good luck, son." I remember SSGT Cumbi to be a good man and I was in no position to shoot the messenger anyway. I cold-sweated the news as the laughter continued. Someone said "why not Minot." It was the first time I heard it, and hardly ever the last till this day. Whatever the hell the last two declared assignments were for the last two fortunate or unfortunate airmen/souls, I can't tell you. It was all about me from that moment forward. I'm impressed that I can tell you that I was third from the last. I suppose some regard for others still filtered through that life changing moment. I was distraught and I needed a map.

There was a globe in the barrack's dayroom. It was the first geographical reference I knew to find. I spun it and found North Dakota. Then I found Minot. Minot had made the globe along with Bismarck and Fargo. This much sort of made me feel better, but what did that mean? Then there was the grave distance from home to consider. As much as I'd always dreamed until that day to see the world as I served my country, there's the obligatory home sickness that we all experience after the aggressive and conditioning experiences of basic training and tech school (to describe it all very mildly), and like any damn kid, I wanted to go home more than ever in my life. In fact, if a CONUS assignment (state-side that is) was to be my fate, the closer to home the better was all I could think. Airman Basic Shane Gerry then walked into the dayroom with a pizza, offered me a slice, and said, "Why not Minot, eh? You know there's like absolutely nobody in North Dakota, right? Ever seen 'Northern Exposure?' Dude, pack warm. That's where you're going."

"There's gotta be somebody there," I replied. "Somebody's gotta pick me up at the airport."

"Well, he won't be hard to find, so there's that," replied Airman Gerry. Airman Gerry was heading to Travis Air Force Base, California, so as friendly as he was, he could shove it up his ass.

I wallowed in self pity for days. SSGT Mejia, one of my tech school instructors, began to notice. Of course the "why not Minot" jabs, along with another developing expression, "freezin's the reason" accosted and plagued me wherever I found myself, so my pain wasn't too difficult to notice. One morning during smoke brake, SSGT Mejia stepped outside and offered me an atlas. It was her atlas and a very nice one. She told me to borrow it overnight and check out North Dakota. She then told me to cheer up. No assignment is ever *that bad,* and it's all what you make of it. For the remainder of tech school, I always felt like she was keeping an eye on me. I've never forgotten her for that.

At the end of the duty day, we all marched back to the barracks, and I couldn't bolt to my room fast enough. I found North Dakota in the atlas index, and the geographical reveals ensued. There was Minot about central state, yet north, close to the Canadian border. There was

Bismarck, far across the state to the south. There was Fargo, also far away, to the east. There were long highways to connect them. I saw an occasional body of water here and there, but they were sparse and small to my standards. There was this rather meek dot declaring itself Dickinson, and another dot you really had to look for calling itself Devil's Lake, and that was literally it. The rest was entirely open to my incredulous speculation. Complete and detailed atlas she said? Seriously, this can't be true. I decided to find home for comparison.

I referred to New York State, and there it was in all its familiar glory. Rochester, New York suddenly looked like a metropolis. Syracuse and Albany and Buffalo were no slouches themselves. Binghamton gave me a friendly wave, and The Finger Lakes and Lake Ontario, and the Adirondacks and the Catskills, and all these big counties and little towns proudly declared themselves for as much as the state could stuff them in. Never in my life had I thought of home as a hopping and happening land of fun and opportunity. Seriously, where in the living hell were the women going to be in Minot, North Dakota? Was I gonna need a tractor? A combine? I'd heard that word before. Is that a thing? Where would I meet chicks? The grain elevator? I wanted to cry, so I shut the book and went to the mess hall instead. Later, Airman Gerry and a few others showed up with beer and Mad Dog 20/20, and some other dirt-cheap vessels of grain alcohol, and as Bukowski once said, I got drunk as a shit skunk in purgatory.

Come the morning, 6am formation never arrived worse. I marched and sweated and farted my way to sobriety and tech school, all the while knowing that every day was a day closer to Minot. I was going to end up a drunk like my grandfather, and Hell was going to be cold. An assignment is what you make of it? Whatever, this shit would piss off the Pope.

I returned the atlas the SSGT Mejia. My hungover demeanor probably said it all, but she didn't mention it.

"Doesn't look like much, does it," she asked.

"No, sergeant, it doesn't," I replied.

"Have you made friends here, Airman Nelson," she asked.

"Yes, ma'am, I've made a few."

"And you'll make them again when you get there. You'll all be in the same boat. You'll figure it out. That's what happens."

And she was right. That's what happened among many happenings throughout my long stay in Minot, but that's another story, and a better one.

I walked away from that chat thinking to myself what Airman Gerry had mentioned the night before. Orders can be traded. It's allowed if you can manage it. It's a long shot, but maybe someone wants to go to Minot? Someone? Anyone? Maybe there's someone around here from North Dakota? Maybe someone wants to go home? Like I said, home sickness is a disease everyone shares by the time they've endured basic training and tech school, and it very seriously plays with your heads. Where is the airman that wants to go home, or at least manage the proximity of home? Can I find that airman? I had to find that airman or die trying.

I never found that airman, but there was a sketchy airman that the powers that be were hounding. He was under fire for brandishing a lot of questionable and controversial tattoos. I won't get into tattoos and the military. Let's just say that you're not supposed to have them? At the very least, you're not supposed to have tattoos that you can't conceal while in uniform, but that regulation was debunked for me right out of the gates. One of my drill sergeants had a forearm tattoo, a rather ornate one, and he was a "blue rope." I promise to explain "blue rope" at another time, but among drill sergeants, it's a big deal. Anyway, this airman was riddled with tattoos, and they were pretty freaky tattoos. This was 1992. Tattoos are about as common as those without tattoos these days, and he had tattoos down his arms, across his chest, and even on his knuckles. The tattoos on his knuckles were not only on his knuckles, but let's just say that they were provocative tattoos. They expressed some rather bad shit. They expressed such rather bad shit that you had to ask yourself how this airman ever made it through a recruiter, let alone basic training and this far into tech school without some stripe of authority asking what the living hell is that? He'd gotten that far and it finally happened. Perhaps he was my first lesson in federal bureaucracy, but I digress. He was under fire for not only the tattoos, but for being an all-around asshole who continued to question authority. He was a sitting duck waiting to be discharged, and he had orders to Lakenheath, Great Britain. My loins called and killed my homesickness. I wanted his orders if I could have them. I went to SSGT Cumbi and begged.

"Airman Nelson, you're going to Minot. Get over it," he replied.

"Sergeant, there's a need for a dental technician in Lakenheath, right," I questioned.

"And they'll get their airman," responded SSGT Cumbi. "Just not you. His orders will be terminated and regenerated. That's how it works."

"I can't solve their problem, Sergeant? I'm ready to go to Lakenheath. Why can't I go to Lakenheath? They need a man. I'm here."

"And then Minot needs an airman. You're their airman. Sorry, Airman Nelson. Get over it. Minot it is."

SSGT Cumbi sympathized. He did. Maybe that day was my second lesson in federal bureaucracy, but truth be told, little as I knew, I would end up better for it. I don't necessarily know where I stand on fate (to this day), but I know I was better off for what bureaucratically happened to me that day, and the life I've lived ever since. Minot was inescapable. It was.

Spoiler alert: I got shit-faced that night, and awakened the next day the same disillusioned airman. The better news at the time was that I would go home before facing the music. I would go home, see my mother and family, get drunk with high school friends, screw my old girlfriend, and don my blues before the plane. It was all overdue and desperately needed, and felt like the only way to go down swinging.

I found myself in my blues at Rochester International Airport. My future awaited in Minot. Best friends (and twins), Brian and Chris Thomas, gave me a filled flask for my sorrows. I drank it all the way to Minneapolis, and grabbed the awaiting crop duster to Minot. The crop duster that brought me there is a story in itself for another time. I got there, and a sole airman (my sponsor) awaited me in a Toyota Celica. It was cold, god damn cold, when I arrived. The rest is another story, and a far better one.

Honestly, when I think of it, I can't even begin.

Make No Mistake

By Lee Nelson

You never miss Mom or anyone or anything you ever took for granted like you do in basic training. You just don't. I'd had life experiences that were difficult and ardent and even impossible, but none of them offered quite the same adversities as basic training. Imagine Limbo or Perdition or even Inferno. You faint to its demands. At some point, you break. You surely do. Any naysayer is a liar.

Basic Training is one of those life changing trials only understood through the experience. It's like child rearing or back pain or the loss of a parent, or being swallowed by a whale. Even with advice you have no idea, and those with the advice know this damn well. Having said that, here's my advice:

Never write a letter in basic training until time has been sanctioned to write a letter. In fact, never do anything at all until allowed to. The truth is basic training is a comedy and a tragedy of errors.

They live for the errors and probably even the comedy. Do nothing at all until told, and once told, own your order. I don't care if they ask you to recite "The Charge of the Light Brigade" while humiliatingly tap dancing to "John Jacob Jingle Heimer Schmidt." Own your order. Absurd requests are fine. Get used to them. Absurd is the word, especially when you're folding your underwear to regulation, or drying the latrine sinks of all their water. Yeah, shit you not: bone dry sinks! They demanded them! They're looking for heart and determination and obedience. In their eyes, failure and success are probably equal. They want growth. They want a progressive measurement of assimilation and nothing more. Ever been to the circus? Dance like the monkeys. Tame like the lions. Independent thought comes later if ever.

My thoughts were clear, exceedingly clear, in fact Tom Cruise and Jack Nicholson "crystal clear" in a letter I wrote to my mother on roundabout day three of basic training. TSGT Burnham found and read those crystal clear thoughts during a surprise inspection. We would all learn there was no other form of inspection. He read my letter while I stood at the position of attention. The embarrassment was bad. The fear of what he'd do next was worse.

He could've read my letter to the whole squadron. He didn't. I'll give him that. Instead, he stripped my hospital cornered bed of all its linens, and tossed them like a father tosses life and hope and belief from your bones and soul, all in service to realities you'll reckon in time. Be thankful. You're welcome. Then he tossed the mattress and bedspring like The Hulk smashes for Marvel fans. Airmen in proximity stood at attention. The rest scattered like cockroaches in the light. I maintained military bearing. If I was to save my ass, this much is rule number one and better be second nature by day three. I awaited his wrath like the condemned. I'd never been condemned. It was an overrated experience.

He demanded pushups. I did them to failure. He demanded sit-ups. I did sit-ups until my guts were knots I'd never learned in the Boy Scouts. Then he ordered squat thrusts. Those are the worst. Enough of those and you'll confess lies or even truths. I did them to failure as well, and he gazed at my pathetic, sweaty heap and called it. Next was the bird shit.

He ordered me to scrub the bird shit from the concrete pad, the formation pad where we always assembled like lightening at any ardent order to assemble. Most assemblies were demanded at 5am. You jumped from your dicks and dreams and landed shiny government issued leather in confused unison, and awaited further screams and shouts. I scrubbed bird shit from that concrete pad with my toothbrush and canteen water for indefinite hours in the southern Texas sun under watch of a fellow airman. It sucked to be either of us. Lots of bird shit, lots of heat. No small talk.

A steel door swung open. Burnham was back. Where were the birds? I could've used more bird shit.

An enraged ten-hut and a march it was. He marched me in the red flag heat to the end of the squadron, and god damn could that man march. I was sweating my body weight across the steaming asphalt of the sweltering sunset. So was he, only he would march with me and then dance around me, then march then dance, all while shouting hate toward man, one man (me), his sweat seeping from the perimeter of his head down the stretch of his campaign cover and dropping on my face as he screamed guttural disdain for me, the lone author of Squadron 3703, a boy who had no business in his beloved Air Force, a boy who could not follow simple directions in a high stress environment, a boy who would surely one day kill his fellow freedom fighting countrymen with his self-serving indifference to order and instruction and disciplined adherence to chain of command! A boy that would surely fail attention to details, details that mean the difference between life and death and defeat and victory, and the fall the of the greatest free country man had ever made! My fathers looked down on me in disgust!

I didn't deserve to serve their sacrifice, their selfless carnage for the freedom of man, their genius design for a brave, self governed society! I was weak! I was selfish! A boy like me kills us all! A boy like me doesn't deserve the honor of the uniform! A boy like me should quit and quit now before he wreaks disaster on the few willing brave who would die for mother and country!

"Airman halt," he ordered. "Step stop," I instructed myself. When ordered to halt, you step once more and stop. Own your order, I'm telling you.

He rounded in front of me and pressed the sweaty brim of his campaign cover into my forehead, and stared into my eyes with disdain. He was exhausted and his breath was powerful. He then shot his right hand in quick salute fashion toward my sternum. He did it so fast I nearly lost my military bearing and braced for impact. I did of course flinch, and with that flinch I found my letter to my mother in a stamped envelope pressed before my eyeballs beneath his sweaty brim. A mailbox was right behind me. Who knew? I sure didn't.

"You wrote it, you fuckin pond scum, now mail it! I'll give you a fuckin head game! I'll give you Hell on earth, Airman Nelson," said TSGT Burnham, quoting my words to my mother. "Stand at attention, God damn you!"

I thought I was. I tried to own my order. He stood at ease, and for a moment, broke Air Force regulation 35-10. Well, it was 35-10 at the time. I've heard the regulation has changed. He removed his campaign cover as we stood outdoors and pulled a handkerchief from the pants pocket of his battle dress uniform. He wiped the sill, put it back on his head and addressed me.

"Airman Nelson, you got two free ones today," he roared. "A pardon and a lesson. Any idea what they were?"

"Sir, Airman Nelson reports as ordered..."

I hadn't the foggiest idea what the living hell he was talking about, and what was left of my melted brain was shear fear.

"No sir," I replied.

"Mommas don't serve, Airmen Nelson, " TSGT Burhnam began. "They've served enough. But mommas should git letters from those who serve. Your pardon ain't for you. It's for your momma. What's the lesson, Airman Nelson?"

Ho boy.

"Sir, Airman Nelson reports as ordered. I don't know, sir."

"The lesson is unwritten rules, Airmen Nelson. Unwritten rules that my beloved Air Force ain't got no regulations for cause anyone who serves knows 'em when they're servin. Nobody survives without writin' their lifelines to those that are right side up when they're upside down. Nobody survived

World War One, World War Two, Korea, Vietnam, and nobody's gonna survive this war on our hands now without writin' letters to their momma, and the right side up world, and if you don't survive, you done right with a letter, a letter of legacy for your loved ones. Letters are for soldiers, not boys, Airman Nelson. I must ask you, do you wanna die?"

That one was easy.

"Sir, Airman Nelson reports as ordered, no sir!"

"Do you wanna be a man that can serve his beloved country?"

"Sir, Airman Nelson reports as ordered, yes sir!"

He relaxed a touch and he looked at me for the first time as if I actually mattered, and asked me...

"Airman Nelson, what must you now do?"

This one I had to get right on the first try, but no pressure, no pressure at all.

"Sir, Airman Nelson reports as ordered. I have to earn the right to exercise unwritten rules."

That had to be right. I had to be a complete loss to the entire system if that wasn't the only answer. But I pushed the envelope.

"Sir, Airmen Nelson reports..."

"Enough with that shit, Airman Nelson, speak your mind," he roared.

"Sergeant Burhnam, I will do as told! The unwritten rules are for those that earn them! I'll learn the details too, TSGT Burnham!"

That's all the gas I had. Trust me you've never heard a squeakier voice in your life.

"That'd be a start, Airman Nelson, a start," replied TSGT Burnham.

And just like that, as quick as the steel door swung open, as quick as formation at 5am, as quick as any shout ever severed an independent thought I dared to ponder in San Antonio Texas, I owned my order to about face and march back across Squadron 3703 to our barracks, our home and its concrete pad, and my canteen and my toothbrush and the bird shit that awaited my return to duty. TSGT Burnham had me grab it all and get the hell to the mess hall. I'd missed dinner. To my experience, nobody misses a meal. Our military is big on that. I have to say it's a good standard.

It took a long time for TSGT Burnham's sermon to have any true impact on me, but it did. It most certainly did. It served its primary directive right away and got an eighteen year old airman basic to discern what's expected of him, and reckon the luck of a free pass he'd most assuredly never receive again. But, ya know, the rest comes later with experience and wisdom. Really, there are obvious reasons here that I've never forgotten TSGT Burnham and have chosen to tell this little story. Scrubbing bird shit indefinitely allows a lot of time for reflection, and TSGT Burnham probably taught me more that evening about serving mother and country than the rest of my four years would ever present let alone reinforce. And though I luckily and thankfully never saw war, never had to bear imperative actions in the face of peril, never truly reckoned the camaraderie that develops in such peril, never bore witness to loss or victory or stood the declared line between "freedom and tyranny," or any declared line between us and them that changes the brave if not kills the brave...

I did undoubtedly learned a few lessons, lessons that have since served my humanity and my maturity, my personal growth and evolution from boy to man with a blip of an idea a minority of people reckon in a lifetime: soldier. I chose to serve my country expecting not to die and to get something out of it. I didn't die and I got a lot out of it. I got a lifetime of way useful stuff out of it, and that evening with TSGT Burnham was a rather large note within that lot.

At the age of eighteen, TSGT Burham taught me a lesson most of us learn if we're lucky enough to live long enough to learn it: loss can be equal to victory. Success teaches the best of lessons. Failure teaches the best of lessons. I received Lesson 1-A in loss and glory via an unsanctioned letter, an angry and committed drill sergeant, and a whole lot of bird shit. It all just took time to sink in to a thick, young head. If you join the military and are worth your weight in salt, you'll learn a few painfully useful things yourself.

Regardless, never do anything in basic training until told or allowed, and own your order. Own your god damn order, I'm telling you. This is my advice, but again, you have no idea. Do not expect a clue, not a one. Good luck, and thanks in advance for your service.

Jeep Rider's Disease

By Lee Nelson

Key:
Amn (Airman)
A1C (Airman First Class)
SRA (Senior Airman)
SSGT (Staff Sergeant)
TSGT (Technical Sergeant)
MSGT (Master Sergeant)
SMSGT (Senior Master Sergeant)
MP (military police)
BDU (Battle Dress Uniform)
AFB (Air Force Base)
35-10 (Air Force regulation citing proper appearance when on duty)

It's very disconcerting to a young man when he discovers he has two assholes. It is. I speak from experience. It was 1993. I was a young airman stationed at Minot Air Force Base, North Dakota, and let's just say something was absolutely not right, so I had to get to the "bottom of the problem," and that required some rather unprecedented mirror work. Yes, you've heard enough, but I have to admit the visualization of the problem gave me pause. I'd never known anyone who'd had two assholes. Why me? It felt like something that might be beyond the realm of military medicine, but what did I know? Yes, a doctor. I needed to see a doctor, so me and my two assholes made our way to Sick Call.

Yes, Sick Call. It's how illness is done in the military. The arrangement is as old as the mess hall, and equally disappointing. One's for the hungry, the other is for the sick, and both are the same cattle call. When you pledge your body to the ownership of the United States Military, you lose a few things, and one of them is independent thought. While you may continue to decide when you're hungry, sickness is no longer a discernable condition, so you have to go to a doctor to make that decision for you. This means that anyone with the nerve to feel unwell is waiting with you in every chair of the waiting room. The chairs are always filled, and almost everyone looks like death. Flu on parade. If you weren't sick before, you are now.

By the way, in 1993, things were a bit different in medicine, both in and out of the military. The culture of privacy had a lot of evolving to do. HIPPA is a commonly understood term these days (Health Insurance Portability and Accountability Act). HIPPA went into effect in 1996, and I can tell you its absence was rather noticeable in 1993. Many in medicine were a big catalyst for why we have it now. Many providers were insensitive to recognizing why anyone wouldn't want to chat up their ailments in the company of an entire waiting room of immature airmen (just like me), or any group of strangers, and in the environment of Sick Call there was no pussyfooting. They attacked this detail head on, like loading trucks. I get it. Nobody ever wants to sacrifice a lunch, including me.

And so it went for all the sick to hear:

"Airman Nelson?"

"Yes ma'am."

"Why are you here?"

"I need to see a doctor, ma'am."

"Okay. Why do you need to see a doctor?"

"I need to have something looked at."

"And what would that be?"

"I'd prefer to tell him, ma'am."

You can see where all this was going. Perhaps I should've just shown her my two assholes, rather than look like I had the clap or crabs leaving my crotch and habituating the chair. Really, that sounds better right now. Maybe I should've done that.

"Can you sum it up for me?"

"Something doesn't look right, ma'am. I'd like an opinion."

After a long stare, she finally picked up what I was laying down and told me to wait to be called, and what a fun wait it was. Yep. Look! There's the guy with the cancer herpes! Stay away from that guy! It might be contagious! I'm a nurse these days. Suffice it to say I've never interviewed like this.

When I finally made it to the doctor, he appeared to have been forewarned.

"So, what's all the fuss about," he enquired.

"No fuss, sir. Something just doesn't look right to me," I answered.

"And what would that be?"

"Well…Sir, umm….is it possible for a man to have two buttholes?"

He wasn't a pediatrician and I wasn't a child, and I'd just asked him if I had two assholes. I'm pretty sure he wasn't prepared for that, but he handled it matter-of-factly.

"Well, anything's possible my friend, but I'm pretty sure you don't have two buttholes. How 'bout I take a look?"

I couldn't drop my pants fast enough. Look at my ass! Please! He wasn't long and he wasn't terribly intrusive.

"Yep. Yeah, you got it. Very common. Happens all the time, unfortunately," was the first he had to say.

I have it? Happens all the time? Please continue. Really, no hurry.

"It's called a pilonidal cyst. You're going to need a minor surgery to fix it, airman, but everything will be fine. It's historically called 'Jeep Rider's Disease.' It was first discovered in World War Two. Lots of soldiers got it from bouncing around in jeeps. It usually starts with a hair follicle that gets tossed around with movement and opens up the pore. It gets bigger and bigger, and stuff gets caught in it. Should I continue?"

I was comfortable with the gist and inquired about the "minor surgery."

"You're gonna go see Dr. Trotter. You'll go to sleep, he'll cut it out, and you'll just have to take it easy for a while."

And that was that. There's no relief in the world like knowing you don't have two assholes, but the condition would prove to have consequences, beginning with my exit through the waiting room. I'd become rather notorious by then, and my notoriety was just warming up.

I mentioned I've been a nurse for a while. That all began with the arbitrary decision of the United States Air Force declaring me a dental technician and hygienist. I must say I never saw that coming in 1992, but as I'm getting comfortable with saying, my life has been better off for it, and that's another story. Anyway, I worked directly upstairs from Sick

Call in the dental clinic of 5th Medical Group, right there in the hospital. It wasn't a large base, and the hospital was to scale. News travels fast. I hardly made it to lunch before my assholes took a life of their own. Did I mention HIPAA?

But there were other rumor worthy developments going on at the time. There always are. One of them was kind of a big deal.

The military was mobilized for a little noted mission in the history books in 1993. It was called "Operation Safe Haven and Safe Passage." To my fledgling enlistment, it appeared to be a defining moment. What might we be called to answer? The 5th Medical Group was decreed to be all in for this mission, and all of us were placed on notice of deployment with our air transportable hospital. The destination was Panama. There were droves of us young and impressionable airmen. It was compelling news to say the least. This was what we signed up for, right? It sure was what we were told we'd signed up for. "We'd pledged our swinging dicks, and you better answer while you're young, dumb and full of cum," to quote a former drill sergeant. Not very original but, it doesn't get plainer than that and obligation appeared to be knocking.

And in this spirit, maybe here is where I should offer some personal background. With the exception of my paternal grandfather, I didn't come from a military family. He'd served in the Navy during the Second World War. The rest found their destinies elsewhere. Probably none of this means anything whatsoever, but I'd committed my entirety to the U.S. military. I'd committed what they wanted, the commodity of a healthy human body, and somewhere along the line they'd convinced me I was nobly serving my country. Must've been the propaganda that accompanied the iron fist of basic training, but in the moment of heeded call, I felt like that should mean something. Would I be ordered to respond to this mobilization? Me and my two assholes were dying to know.

And so the call came and initial excitement ensued, but then nothing much developed, so we waited. We went on with our normal duties and we waited. Then we did some waiting. I suppose unless a bomb drops at your door, bureaucracy wins the day every time. Information wasn't disseminating to the minions. However, "recall drills" were on the rise.

Recall is essentially an emergency response drill, or a war time response drill. The expectation is simple: get your ass to work right now! Oh, and get your ass (or two asses) to work in BDUs. For us medical personnel, this meant the difference between arriving in camouflage uniforms or our time-stamped and traditional medical white uniforms. The whites were really bad, by the way, and I only hope they've been improved or phased out. But I digress, and I'll tell ya you were never at a loss to find the sole airman that arrived to work in medical whites and zero clue of the recall. It can happen to anyone, and it's pretty embarrassing. It doesn't go without insult, both in ridicule from your peers, and chastisement from your chain of command. I got through four years without that experience. Sometimes the mistakes of others are all you need. But I made plenty of mistakes myself, so don't let me fool you.

Anyway, back to recall drills. All peacetime duties are continued but minimized, and percentages of personnel are "activated" to respond to the war time/emergency mission, whatever it may be. You're always the last to know why you're responding, but you need to respond immediately and without question. Don't bother to shit, shower, shave, brush your teeth, comb your hair, nothing. Do absolutely nothing whatsoever that could impede your timely response. It's simple: in peacetime, look like a million bucks. In wartime, the worse you look when you show up, the better. Looking like shit wins the moment. Even smelling bad is fine. It's the smell of an ardent soldier smelling bad but serving well. You can see where all this makes arriving in your well pressed whites even worse. Good! Now you understand the embarrassment. If you ever join the military, get the recall etiquette down immediately. You're welcome.

So, you get the recall notification from a pyramid of personnel phone calling. Everyone gets a copy of the pyramid. The calling pyramid goes from the brass all the way down to the airmen. I always kept a copy of the pyramid on the fridge. You get the call, respond, and call the next airman on the list. If you don't reach this airman, you call the next one, and sucks to be the airman that didn't answer the phone. If you're single and living in the base dorms, you'll hear a knock on your door, or at least you better hear it, and so it goes. The mob of the stinky and disheveled ensues. Cars and buses of stinky soldiers cross this way and that at any declared hour as all hands report to work. Once at work, you find out what's next, but never without bureaucracy. "Hurry up and wait" is the game. Once told what to do, whatever it is it sucks. You can either be told to resume normal duties, which means doing your job all day smelling like crap, or you're told you're getting "activated," and man you better be prepared to be activated. Your end of getting activated isn't a whole lot to ask. God help you if you aren't prepared.

What does it mean to be activated? Glad you asked. By the way, I'd venture to guess this is all handled with cellular alerts these days, but anywho....

Everyone must have a "deployment bag" prepared in their quarters. Your deployment bag is the military issued duffel bag everyone receives in basic training. There's a lot of room in that bag, and it's quite durable and can sustain a great deal of weight. This you handily learn in basic training, but anyway, the deployment bag. It's commonly abbreviated to D-bag, and man did I love D-bag. No nickname better suited the bureaucracy and organized confusion. You simply had to have in your D-bag several BDUs, an extra hat, pairs of underwear, socks and toiletries, and money and other pertinences if you ever gave any thought to leave the country on a second's notice. The D-bag must sit ready and unmolested at all times. This was everyone's end of the deal in the minimal sense. Not much to screw up here, is it? Well, you'd be surprised. There's always "that guy," and that guy was once me (but only once!), but thankfully, I got through my unpreparedness with little notice.

The recalls went on and on. With every recall, I found myself returned to duty looking and smelling like crap. Others were called to respond, but never me. Young airmen get lazy, and one of the challenges of maintaining a D-bag is the constant need for clean BDUs. You only have so many to go around, and who wants to do laundry every single night? I sure didn't, and my ex-wife and I had a young son, and ya know life goes on. Stuff just gets in the way, but it cannot get in the way. Nothing in your life can get in the way of a military mission, not ever, not even two assholes. The military always comes first, and you damn well better know this up front. Your family is not the priority. You are not the priority. The military is the one and only priority. Any questions? No, there are not.

Anyway, for all this truth, still I make excuses. My unprepared D-bag was on me when I was finally chosen to be activated. I knew it too. I just knew it was going to happen as I was recalling myself (how's that for a term?) with an unprepared D-bag waiting to mock me. My ex-wife had even warned me the previous evening. "You better get your bag back together. You're gonna regret it," she told me. I didn't heed her advice. I sat and drank beer and watched Seinfeld if I remember correctly. I could've used more beer in the morning.

"AMN Nelson," said SSGT Hargrove.

"Yes, Sergeant," I responded.

"Go get your D-bag. You are activated."

"Yes, Sergeant."

Son of a bitch! I was wearing my only clean uniform. I blazed through the hospital and out to my car knowing what I had to pathetically do. The hospital was outside of the base, and I remember approaching the guard gate so fast that the MP at the gate gave me shit about it. I zoomed past him and to home at stupid speed and pulled all my dirty, stinky BDUs out of the hamper and placed them in my D-bag. I might be forgetting a few other stupid details, but I do recall heading for the door and realizing I'd robbed my toothbrush from the D-bag several days before. My last toothbrush had lived its lifespan. I was a dental technician, remember? "Can't get caught without a toothbrush," I thought.

I went back for the toothbrush and suddenly felt something strange, something rather unpleasant. Sorry for the ensuing details, but it's the story. My second ass suddenly required address. My abatement measures had failed at the worst moment, and emergency readdress was required. Yeah, that's enough of that. I got myself cleaned up, jimmy-rigged a second-ass abatement device, and out the door I went. This of course required time I didn't have. I left the house in a heap, raced back to the hospital with D-bag in hand and in fear of failure of my jimmied second-asshole dressing. My untimely arrival was noted.

"Where ya been, AMN Nelson," asked SSGT Hargrove.

"I've got my D-bag, Sergeant," I responded.

"Did you move to Zimbabwe," asked TSGT Hopkins.

For the record, TSGT George Hopkins was always the ball buster and I grew to love him for it, just not right then when I was sporting a disturbing condition and guilty of whatever he decided. Oh, and I've maintained a fun social media relationship with him for a long time now, and he prefers to be addressed as "The Awesome and Majestic TSGT Hopkins." He's a damn good man and he taught me a lot, but I don't know about the title. I do often wonder how he carries himself these days. He was always the gate keeper of Air Force regulation 35-10.

"I did not, Sergeant Hopkins."

"Whatever," said SSGT Hargrove. "The bus is up front. Get to the main entrance right away."

"Roger, Sergeant."

"I like your lack of 35-10, AMN Nelson. Your country is running out of patience," said TSGT Hopkins as I ran for the door. If I had looked the way I did under any other circumstances, TSGT Hopkins would've eaten my face for breakfast.

I made the bus. It literally stopped just for me. I schlepped my D-bag of stinky uniforms to the bus's underbelly and hopped on. Walking down the aisle, my underwear harbored abatement measures began to argue with me. "Please, God," I thought to myself. "I'm really struggling here." I found my former roommate and good friend to this day AMN Danny Burd, and sat down next to him and sighed.

"Nelson, you're a mess," said Danny.

"Yes I am," I replied. "You've done this before. What happens now?"

"Remember when I told you we drove around the base for an hour and then we sat in front of the hangers for another two?"

"Yeah. That's what we're doing?"

"How the fuck should I know? That'd be my guess. We'll see what happens."

And we drove around the base for at least an hour or maybe the act of sitting on a bus driving aimlessly to every corner of the base and back made it feel that way. Our best guess was that this was a simulated excursion to wherever the hell we were going, which was Panama? I really didn't want to take a Greyhound to Panama. Nobody did. The idea was as absurd as two assholes, but we eventually arrived at the flight line hangers staring in awe at B-52 bombers up close. By the way, it wasn't a Greyhound. Greyhounds are in better shape.

And then we sat on the bus in front of the hangers and B-52s until the return of Christ. Digestive and urinary complaints began to mushroom among the airmen. It was getting a little desperate and then...well, you probably know what then.

"Doin all right over there," asked AMN Burd.

"It's noticeable," I asked him.

"Nah, but you're noticeable. You keep squirmin."

And I kept squirming.

"Havin a bad asshole day," AMN Burd inquired.

"Umm, yes. I gotta get out of here. Something's gonna happen and I'll have to live with it the rest of my life," I replied.

"When do you get that ass fixed?"

"I see Major Trotter in two days. Can't wait."

"I hear he charges $500 a crack."

"Fuck you."

"Maybe you can see his partner instead: Captain Seymour Cheeks."

"Fuck you. I told you how you get this damn problem, didn't I?"

"I didn't need to hear it, but yeah. Yeah you did."

"Well, maybe you won't be laughing after another hour on this god damn bus."

"You got a pile of shit in your D-bag, don't you?"

"Fuck you."

Finally, we were ordered off the bus to schlep our D-bags into a hanger. Further waiting ensued and then we waited. Then we waited some more. But right off the bus I managed a trip to the latrine. Things weren't as bad as I thought. Uncomfortable, but my rigged abatement device was holding. We all smoked a lot of cigarettes and waited. That's when it happened. SRA Wyatt asked how my ass was doing, and that was that. About a half dozen guys had been waiting for the inquiry. Good times. "Airman Assholes" was born and the rest is history. Did I mention HIPAA?

Then some SSGT or TSGT or MSGT ordered us all to place our D-bags in the beds of some big ass pickup trucks and we did. This made the drill feel kind of official, like things were about to get more real than any of us were prepared for, or at least me. I thought about my son and ex-wife. Was I about to disappear? Can I get a phone call? After that, I bet you can guess what happened. Yes, we waited and smoked cigarettes. Then we waited some more. After so many hours of this ridiculousness, a minor, unaddressed detail began to linger. We were all hungry and thirsty, and as for me, what about my asses? Did I mention the military comes first?

"Room Ten Hut," shouted some form of sergeant, perhaps the same one that had ordered all the D-bag schlepping. This is a common sudden notice. An officer had entered the hanger. All enlisted stand at attention. Could be the next Patton or a lost, clueless and hairless boy with bars on his shoulders. Either way, the honorary recognition is the same. Turns out we were happy to see him. He was a second lieutenant. "Butter bars." That means nothing's happening.

He left the hanger. The same sergeant then ordered us all to grab our D-bags and put them back on the buses. That was fun. Blind led the blind through chain-ganged who's whose D-bags back to the underbellies of the buses, buses with drivers that didn't give a shit who you were or what bus you arrived on anymore than the buses did. We didn't drive aimlessly around the base after that. We were brought right back to our squadrons and dismissed, but the D-bag confusion still took a lot of time. It was a damn late hour when my starving body finally got home. My ex-wife understood and my son was happy to see me. Imagine any such happening in a civilian scenario. "Honey, I was possessed indefinitely by work." I suppose the same can happen in many civilian professions, but I've made my point clear. So sayeth the military. The military comes first.

Two days and one recall later, I finally had my appointment with Major Trotter. He may as well have been Jesus. He was the only man on Minot Air Force Base that was going to cure my assholes. I was giddy. He was all business. He was a young guy maybe ten years my senior. He was comfortable with solving my problem. I was all ears.

"So, the surgery isn't much. What's important to know is that this is a long haul recovery," he began. "There's two approaches to the pilonidal cyst. One is to drain the cyst, cut the margins around it and sew up the hole. The other is to do all that and not sew up the hole. I've found the former has a high incidence of the cyst coming back. The latter does not, so I only do the latter and I don't sew it up. This means you're going to have an open wound for a while as it heals from the inside out. It's going to hurt a lot, especially in the beginning, and you're going to need the wound packed twice a day, so it's not a fun time, and you're gonna need someone to change the packing. Any questions?"

"How big of a wound are we talking here, sir?"

"I don't know. Depends on the size of the cyst once I get in there, but it could be big. It can happen."

"And somebody has to pack my ass twice a day?"

"Yes, that's very important."

"Who's gonna do that?"

A short pause ensued and Dr. Trotter looked at me as if I wasn't too intelligent.

"You're married, is that correct," he asked.

"Yes."

"Well, for better or worse, right?"

"Yes sir. Sounds pretty worse to me."

Okay, quick reveal here: I've been keeping my ex-wife angle in this story short and simple, but she's my ex-wife for a reason. It was hardly ever a happy marriage, and things were pretty damn bad at the time. I honestly didn't see ass wound packing in the cards.

"Sir, I'm gonna say that's not an option."

"Well then you'll have to come here twice a day for wound changes. You're right in the hospital anyway. Shouldn't be a problem," he replied.

"And what about weekends," I inquired.

"You'll have to come to the emergency department. I suppose it sounds like a lot, but there's nothing to the dressing changes. You'll be in and out in no time."

Then there was another pause.

"Are you sure your wife isn't an option," asked Dr. Trotter.

I gave it another thought. Option two wasn't sounding like a lot of fun, but then I imagined a woman with a grudge packing my ass twice a day.

"Yeah, I'm afraid she isn't, sir."

"Well, any more questions?"

"No sir."

"Let's get you scheduled."

I then found myself with Major Trotter's secretary, and as I sat there getting scheduled, guys I knew in his office passed by and I imagined them redressing my ass. Then I thought of the couple friends I had in the emergency department redressing my ass. My fate was sealed. Airman Assholes was about to get way epic. Did I mention HIPAA?

As he promised, the surgery wasn't anything, but for the record it was a big god damn wound! Wow was it big. And he was right, the recovery was everything. I wouldn't wish that recovery on my worst enemy. I was given two weeks convalescence with a lot of pain killers, and a trusty inflatable toilet seat. You haven't lived until you walk the earth indefinitely with a toilet seat in your hand. Yep. Here comes Airman Assholes with his no-secret weapon. I tried really hard to live without the toilet seat, but it proved too arduous for this particular soldier. I needed the damn thing. I sure did. Danny Burd came to refer to me as "the Turkish prisoner." Must've been the way I walked. Oh, and of course my friends in the surgery clinic and emergency department got real chummy for a while in a way none of us ever imagined. Yes, I've mentioned HIPAA. Regardless, the embarrassment and jokes subsided. I suppose it's hard not to have sympathy for a man who can't sit on his own ass.

After two weeks, there was no reason I couldn't clean teeth, so back to work I went. The sitting got easier and the wound took me off the activation responsibilities, so I quietly did my job and watched the world recall around me. That was kind of nice, but TSGT Hopkins never missed an opportunity to have fun with it.

"No two assholes ever kept a man out of a foxhole, AMN Nelson."

"Noted, Sergeant."

"How's the keister today? Fit for duty?"

"Better, Sergeant, but if you're here to chew my ass, you're too late. Wanna see?"

"I'll pass. Carry on."

So, the mystery of Operation Safe Haven and Safe Passage continued to linger in the air. The recalls continued as did the threats to all of us being separated from our loved ones. As I mentioned early on, if it were to become a reality for us, I wanted to give all I had to serve my country. That's what it's all about when you make the pledge, right? I most certainly didn't want to be separated from my son, and if there was any chance in the world to save my marriage, months of separation wouldn't be therapeutic, but really, what did I get in this for?

The short and honest answer was to "see the world" and earn my GI Bill, but once you're in and serving, things become far more than that. You can chalk it all up to military conditioning to serve mother and country, and that might be true. I found it to be a question of credence. There's a lot of horse shit in military propaganda, but the question all the while remains: do you believe in your country? If you don't then why are you here? I wanted to do my part, but as you may have surmised by now, my two assholes had other plans. Yes, I was officially on injured reserve. There was zero chance the Air Force was going to send me to a second world country with an open wound, wherever that wound may be. I had been grappling with this since the diagnosis, but I'd chosen not to think about it and see how it all played out. It proved to play out as reason would expect. My mission was home front. Clean the teeth, Airman. I cleaned the teeth and worked on my marriage. That much went well for a while. I got my daughter out of all that temporary goodness, but that's another story and a better one.

Such things didn't go quite as well for many colleagues. The mission was finally realized. Orders and timeline came to light, and there were a lot of meetings. My god was there meetings. There were meetings to discuss the next meeting and plan further meetings, and I was begrudgingly involved in countless meetings despite my two assholes and lack of expectations. Many of us were deploying and all of us were preparing for it. The air transportable hospital (anyone should see the damn thing, it's impressive) was constructed and destructed in a field near the hospital at least twice before the final salutes to whoever among us was leaving. Major Trotter was among the many, so I ended up under the care of his partner. His name wasn't Captain Seymour Cheeks. It was Captain Christopher Moberly.

They all left one day rather anticlimactically. I suppose so because I can't remember much about it. I do, however, remember the impact of their exit. I'll never forget it. It meant a lot to everyone left behind. I suppose I was young and selfish and concerned with my own family and two assholes, but the impact of their leaving was very evident to all of us. Family separations are tough no matter whom you are, but we were young. We were all damn young. Such an event is a lot on young lives and it creates drama anyone would have, but I would argue the drama is only made worse when immature and unworldly. It was ugly and hard to witness. In hindsight, I was glad for my "Jeep Rider's Syndrome." I had enough problems despite my earnestness to serve. Boy did the climate change on Minot AFB, and while I've told a long story already, please allow me to elaborate on this. I feel it's important.

The military enhances a family dynamic that I'm at a loss to know exists otherwise. The military doesn't promote this dynamic. Rather, it's an unavoidable consequence of having millions of young men and women dispersed all around the world. Imagine millions of enamored and horny college aged kids not going to college, but thrust into adult and professional lives. I don't know how to better articulate it, so let's just say that many of us get lost in a perception that we're older than we are and more prepared than we are. A lot of us quickly grab each other and hold. I sure did, and Minot AFB housing was filled to the hilt with exceedingly young families. It's strange to look back on, but there was nothing unusual about eighteen to twenty two year old married couples as far as the eyes could see, and the children arrived quickly. At this time, I'm forty eight, my son is twenty eight, and my daughter is twenty three. Did the military change my life? Probably more so than any choice I've ever made.

You get a good separation of young families and suddenly a great many airmen revert to more age appropriate yet family devastating behaviors. You see it coming but the shock isn't dulled. There were drunken horn-dog fests going on in Panama, and they were going on at home. They were everywhere. A whole lot of us had a great time, but you know every wild party ends with a sobering hangover. Like I said, the drama was ugly and everywhere. I suppose far worse military stories can be told. Nobody died to my recollection, but many young lives were sadly and significantly changed. Families were uprooted. Domestic violence was terribly common, as was depression and alcoholism and suicide. Yes, suicide. So I guess there was carnage after all. Oh, and court-martials. Careers abruptly ended too. I was thankfully immune, but I grew up a lot by default. What a rollercoaster it was. I'd never been through anything quite like it before and haven't since.

I'm proud of how my own family came out of all that, at least for a while. We were as young as any of us were and we had our problems, but my ex-wife and I frankly ended up giving it all a pretty good run. We gave a real shot at "for better or worse." After four years in the military, the question of reenlistment arrived, and it was a damn tough decision. I eventually found where my heart truly lay, and my ex and I decided our course of action. She wanted to reenlist and I didn't. MY GI Bill was burning a hole in my pocket and I wanted to work in the private sector and go to school. We made it work for a while, and all in all we lived on Minot AFB for six years. A lot changed in that time and the seeds of the rest of our lives were sowed. But again, that's another story, most of which is a far better one.

I became a surgical technologist in that time and paved my way to nursing school. Funny note: I ended up working in the private sector with Doctor Lee Trotter, and we had my ass and work in common for a few more years. And I made more friends. I'll surely never forget Minot.

I'm a better man for the military, but the military comes first. What does that mean? I hope I've touched that a bit. Not every military story is a commonly told story, and they all matter. Do I offer any advice? Of course I do: minimize your time in jeeps and trucks on bumpy roads. Keep a tight D-bag. Bureaucracy will win most days, and duty may or may not call. This is as arbitrary as service will always be, so it isn't on you. Do all you can to love your family first under the warm dragon's wing of mother and country, and wash your god damn uniforms.

As long as you do your duty, there's no shame in what may happen to you, not even two assholes. Oh, and whoever you are, wherever you are, military or not, be thankful for HIPAA. It's a good thing. It matters. It helps.

I've been an operating room nurse four almost twenty four years now. It's amazing where a GI bill and two assholes can lead a man. I'm often asked if I'm a veteran. The question comes from discerned work ethic and demeanor. It's always a compliment. It always nurtures relationships. I'll always be thankful for my service. I'm at a loss for anything that's been more positively impactful to the compass of my life ever since.

Sad Story

By Lee Nelson

I have no fond memories of Msgt Jane Saunders. I would tell you to Google her, but I've given her an alias. She made headlines for terribly unfortunate reasons, reasons I wouldn't wish on a worst enemy in the world regardless of what you are about to hear. I have no fond memories of MSGT Jane Saunders. She may actually be the singular reason I chose not to reenlist and walked away from the military forever.

The military is filled with good people. They're everywhere in countless numbers, but truth be told, even the worst among them are noble. They're serving their country. Where the rubber meets the road, even the bastards are the best as Jack Nicholson's Colonel Nathan Jessup said it best: "You want me on that wall. You need me on that wall." They serve a detrimental service, no matter their character or lack thereof.

I have to admit that Msgt Saunders had character. She did indeed. She just didn't have character anyone could appreciate, particularly her subordinates. Perhaps in basic training she would've been useful, even exceedingly useful. She was committed to the iron fist of authority only basic training can best utilize. She worked in fear and degradation and dominance. It appeared she'd learned long before me that somehow the best of performance can be born from hostile and overbearing supervision. I never ascribed to that philosophy, even in basic training, and even those that perhaps did ascribe to that philosophy didn't dig her approach anymore than I did. She was a son of a bitch. She abused her authority with what appeared to be pleasure, and these are about the kindest words I'll ever manage. I could give you several examples, but I'm far too susceptible to digression. If an alias wasn't so important I'd tell you once again Google her. Her unfortunate story speaks for itself.

To this day, I'm convinced she sought a recognition she could never obtain. Perhaps it was something as simple as a salute, something she would never customarily receive that we would hand out to the officers like you would any boss; an obligatory and empty gesture of respect to, on average, a passing stranger. She was noncommissioned. She never got that. This is my speculation, but it sticks with me. I'm otherwise at a complete loss to explain this human being.

I had experienced a lesser form of Msgt Saunders in a previous Msgt once upon a time in four years of service. I was young, but my gut wouldn't let go of a few nagging questions: how many of these bastards will I meet in a career? Do they exist outside of the military? Probably, but to what extent could any such person exercise power over me if I were to leave the Air Force? It's a sad ending (if that's the ending, but I honestly don't know to this day), but she was unbearable in an immediate and foreseeable sense. She sold me on my walking papers, and that much is very sad.

And there is another tragedy here: her fallibilities, in the right context, were priceless qualities. Like I said, basic training, but I would imagine such a human serving her country must've had other better applications. I wouldn't know. Perhaps you do, but there's an irony here: she had all the stuff for the best you never forget, and I'll never forget her for all the wrong reasons. The military needs people like her. It's impossible to best utilize all the numbers, and a gem in a pile of shit gets shoveled away. In all the stench, you rid it with fury.

If a gem never finds a ring, it's never a gem, and in the military, it sure as hell never gets a salute.

Jim Sutherland

By Lee Nelson

I was going home. Nothing was going to stop me. It was my first chance, a bird in hand, and I grabbed it. It was sort of like the desire to devour a fat sandwich on cheat day, or drink a beer out of the gates of drying up. Maybe even better. Maybe it was more like the guard gates of prison opening to the highway and the sedan and the one or few loved ones waiting to embrace you. I can't be sure. I've never been to prison, but it was nonetheless comparative freedom. I was hot and sweaty and it was damn cold water. I left on leave from the military for the very first time, and I had wild Spring Break caliber good times with all my high school friends, and it all hit and ran like a jackrabbit on a date. By the end of it all, I still hadn't managed to see my good friend Jim Sutherland, so we met up at a dive bar and drank and shot pool and threw darts before I returned myself to the property of The United States Air Force. It was a good night with a good friend, and a night I'll never forget. But I had no idea why I'd never forget it. The true power of that evening I wouldn't know for another twenty eight years.

Jim Sutherland and I have been friends since elementary school. Out of high school, he did what most of my friends did and went straight to college. If I remember correctly, he hated it. He went the route of community college and stayed home with his parents because he was like me: he couldn't afford much more and wasn't interested in massive college debt. He was on the "two plus two plan" where you start with two years of community college and finish with two years at university. He was doing well enough and getting by part time at a grocery store, but community college was anticlimactic for a college experience, and he was learning along the way what most of us learn: he had no idea what he actually wanted to do. He had no idea and he was spending money he didn't have to figure it out. He was bummed.

I remember him asking me how to know he'd survive it all. How would he know after all that debt he'd come out with a survivable wage? I hadn't thought about any of that. I'd surrendered myself to my country and was happy to get paid, drink beer and chase women. He liked my attitude and was impressed that I'd done something different out of the gates, and he was impressed with my "freedom." After enough shots and beers, he was in a better mood, and the night ensued.

There were shots and beers and darts lousily tossed, and a few rounds of pool that were equally lousy. I remember talking to a couple of women. That didn't go too swell, so we slammed more beer and shots and talked long into the night.

I remember how worried he was about the future, and how serious he was all around compared to my utter lack of seriousness and worry. Having just completed basic training and tech school, perhaps I felt like I had accomplished something, just enough to be content with my foreseeable future? More likely, I was just a dumb kid who didn't give enough of a shit about anything, and I was peppy and drunk and talked a good game. Whatever it was, it left an impression, and Jim asked me a great deal about the military. He had a lot of drunken questions and I drunkenly answered them, or maybe I didn't. I can't begin to tell you what I may or may not have said about my fledgling military experience thus far, but it was all the conversation and he continued to pry and challenge me with questions.

Other than that, I remember he absolutely hated the hoodie I was wearing. It had horizontal purple stripes, and he couldn't stand it and told me to burn it. I loved the damn thing. I wore it for a few more years. Nobody else ever told me it was ugly, so I have the fashion sense and he does not. But I don't know, maybe the purple stripes were the reason we didn't get laid, or maybe it was the countless shots and beers, but anyway, I divulged all I could best tell him for all its shit-faced value.

I can't tell you how the night ended, but it eventually did. I could tell you how the next morning went, but it's no Colombo mystery. I'm just glad I made the plane.

From there, lives took off. Never close your eyes and never forget to pause and smell the proverbial flowers, let alone regard your loved ones or anything else that swells your soul. Jim and I hardly talked for some years after that. I do remember a phone call one day when he told me he was enlisting in the Air Force. I asked him if he was sure. He more or less told me he was.

It's understatedly important to note that his father is a decorated Vietnam marine that served in a capacity neither of us will ever know, but are inexpressibly proud to honor, and thankful we'll never better understand. Regardless of impetus, Jim enlisted. He made the pledge and plunged right in. I did my four years and walked. Jim Sutherland hardly finished comparably.

Jim Sutherland went career military. Jim Sutherland lived an Air Force career that went from Lackland AFB, San Antonio Texas to many places in not necessarily the following order: Charleston SC, Pope AFB NC, Moron Spain, Greenland, Ramstein Germany, Dover Delaware, Sigonella Sicily, Lages Field Azores, Bangor Maine, Spangdalem Germany, Rhein Main Germany, Amsterdam Netherlands, France, Luxemburg, Bargram Afghanistan, Royal Air Force Fairford UK, Greece, Diego Garcia, BIOT Kurdistan, Aviano Italy, Asheville NC, Little Rock Arkansas, Balad Iraq, Shannon Ireland, Kandahar Afghanistan, Al Udied Qatar, Biloxi Mississippi, Kapaun Germany, Stuttgart Germany, Aseleem Kuwait, Eglin Florida, Scott AFB Illinois, Joint Base McGuire-Dix-Lakehurst NJ, and some other locations too classified to mention that he can't admit until the age of seventy, classified contracts being what they are.

He did a lot and went on to do a lot. He retired with the rank of master sergeant and a master's degree, and is still currently serving our country as a contracted civilian.

He has a wife and child. They live comfortably on the home front. He's happy. He found his calling, served his country, and quietly serves us all. He was one man in our mighty military, and it's safe to say he did his duty. We'll never have any idea. That's pretty much how it works for the quiet countless that have served and continue to serve Mother and Country.

Jim Sutherland called me three years ago to announce his retirement, and ask me to be there for the ceremony. I wish I'd had a chance at being there. I didn't. On the morning of his retirement, I called him from my truck on the way to work. The circumstances proved sad and surreal for what he had to say to me.

He told me I'd inspired him to join the military. He told me he wouldn't have if I hadn't convinced him. He said thank you for helping him point his direction, and thank you for being a good friend. Then he donned his blues and left with his family for his ceremony.

What the living hell does anyone say to that? The least I can say is that it moved me in way that stuck with me all day, and has stuck with me ever since. He says I did this for him. Did I? I don't know and I will never know, so I had to share. You be the judge.

Regardless, I can tell you this: don't try too hard to toss a rock in a pond. If something big is going to happen, it's going to happen. Changing someone's life for the better, or any good deed for that matter, can arise inconspicuously, even unknowingly, and can present in the strangest and most common of circumstances.

So, try not to worry any more about things than I didn't that shot filled evening, and try to worry far less than he did. We all find our way. It's all up to us and the only impedance to any of us is freedom. You gotta have freedom. That's the one and only key to all and everything.

Alive and at the Helm?

By Lee Nelson

It will call.
It always calls.
It calls
and there's no such thing as rational
and it called upon a 2am
on a summer night.
I realized I had everything I need.
I've always had all that I need
even the balls to do it.
I crept out of the house
towel in hand
and down the road to
possibilities.
It does help to live near a lake.

You can disappear into any lake
but you've all heard of Ontario:
smallest of the largest we have.
Pitch black
all alone
in stars and silence
I swam past the bars
where only me and my own
could fend chance
and current
and perhaps Loch Ness
but again, Ontario.
It was plenty for the job.

I was
I am

much older
than the hunger that wills
such mission
such power
across dangerous water
despite the hunger
the call
that will always be my years
and I got way out there
to where cramps
and declining stamina
and all wills of what will be
could declare who's in charge
and I stroked, paddled
and floated the surface
of reasonable death
tranquil as it was
and found rare peace.

The abyss of my neighborhood
and the stars of Mother Abyss
reminded me what frailty is
when I thought for years
I already knew
and the silence that accompanied
a distance birds of prey
tap out...
I can't tell you.
I will call the call the call
and I will tell this as this
was medicine.
No doctor will prescribe it
and I will not endorse it
but it was all peace

buoyant surrender to no land
the only place where none of it is
and the most mute direct threat I've ever known
the greatest harrow I've never heard
or had ever welcomed.

I never thought of my children.
I never thought of my wife.
I never thought of any paramount matter
in my small world that truly matters.
I can tell you it was all worth doing.

If I'd gone down there to find
wind and waves or any other
formidable portends of
Mother Nature
I would've skipped rocks for a bit
gone home to pharmacology
and called it a night.

The raw world can rock
a baby to sleep.
This the towel taught me
as it dried my peace
and reminded me
why I brought it.

Vaughn Stelzenmuller

Vaughn Stelzenmuller writes to get people to read. His five-year military career piled up ample material to do just that. After commissioning into the Regular Army from Johns Hopkins University in 1967, the government posted him to Fort Hood, Texas; Central Highlands, Republic of Vietnam; and then Stuttgart, Germany. There were the "usual" or expected Army schools along the way. He writes about what happened as an infantry officer, the sorts of things about which most veterans would jump in with a "Me too" confession. Many of those things can be humorous, some less so. When veterans write, everybody learns. Vaughn made captain after two years' active service—which every officer did at that time, at least the ones who refrained from getting sent to the stockade. After active duty Eastman Kodak Company hired him as a manufacturing engineer. Now he writes. He is still working for "promotion to captain" in real life. It isn't always easy, but always instructive.

"Thank You for Your Service...."

It is great to hear those words these days, though it makes me glad, worried, and hopeful all at once.

Coming from the Vietnam era, speaking that phrase was not fashionable. It is much more so now, though it's pretty much as shallow as years ago, in my irritable opinion. Still, civility and kindness aren't so bad, and I'm truly *glad* to hear it.

What should *worry* everybody is that this general well-wishing to Service people gives ""cover" for our national leaders to meddle—to show off U.S. combat power just for the hell of it, for the sake of making themselves feel important and "good." This leads to the temptation to knee-jerk thoughtless precious tax dollars onto so many military-connected measures, with neglectful oversight.

In particular, *do* recall how a small unit of Army Special Forces came within two whiskers of obliterating Al Qaeda in the 2001 Afghan mountains. And then the full weight of U.S. Army stormed in to "help out." That calculation bought the U.S. another two decades of combat. Clearly, the war party in Washington didn't want to miss this handy, clumsy "party." I rest my case.

Still, I fervently *hope* that we figure out ways to keep our armed forces, *all* of them, in sharp readiness state. Since this is *not* the same thing as testing readiness with foreign armies—"kinetic action" really, as some leaders slink from saying the words out loud—we need smarter leaders to understand the stark difference. This boils down to constantly rethinking and evolving plans and means. This has to be done without constant combat operations for their own sake.

Personally, I do not require the verbal "thanks" for my service time, although I understand others' taking the time to speak those feelings. As for the verbal "curses" in previous decades, I have always ignored them. Service was done for the service itself, and I came out a much more useful person for it. That includes *lessons learned* about constant rethinking and evolving, personally.

Homer the Mouth

Shots!

Every Service person gets shots in the arm. Both arms. Lots of shots. Extras, if the lucky Service person will be assigned to an exotic place, where every germ in the book lurks in the air and water. Even if they really don't. Lucky, lucky Service person.

Long, long, long ago these inoculations were given as a "battery" of needles, all at the same visit. Usually, service people get this battery along with a dozen-gaggle-plus others awaiting the same lucky fate, same T-shirt ready room.

One of the smaller pleasures of military life, though, comes when you go through tedious stuff all by yourself, when an entire battalion isn't watching you. That almost happened to me by really lucky 1968 circumstance, so picture this scene:

> *Here's this new lieutenant, alone in one of the post's medical rooms, awaiting Dr. Army Major Jekyll to walk in and make my arm an iron shield against the world's exotic-place sicknesses, but otherwise less useful for the rest of the day.*

Then: Oh, what Fortune! In walks a small squad of enlisted men to get their deserved shots, Homer the Mouth in the lead. *Mouthpiece* persons, in or out of uniform, are always talking first, loud and inappropriate. *"Ah, here's an officer. He's going to show us how it's done!"* chuckles hotshot Homer to his group.

Why does a *shit-eatin' grin* always come along with individuals like this? In the Army this usually sports the perp as an E4, a *specialist-fourth class* rank more exactly. It's the perfect rank to get away with damn near anything. An E4 *corporal* would know better. New lieutenants can't.

All right, the happy ending to the story. If part of that plastered ceiling just then had fallen on me, and then a large mastiff loped through the door and chomped onto my right arm, along with the doctor tackling me to the floor medically to get my blood flowing, all at the same minute, I would *not* have flinched. Nor flinching for some piddling little serum injection. Five injections, as a matter of fact, in front of a squad of chuckling enlisted men.

I, a staunch no-flinch lieutenant, took the needles without so much as an eye-bat. Then it was deliberately and "in a military manner" smartly putting the *blouse* (Army talk for "shirt") back on. Then out the door without smiling, but with a great bored David Niven nod to the small group. Now it was the turn for *their* collective left arms.

I recalled that Mr. Visnic, our high school biology teacher, had bragged to us that he was made to do 45 minutes of *manual-of-arms* drills after getting shots during his younger Service days, and that kept his arm less achy the next day. Therefore, it was out the door for me, then back to the orderly room to check out an M1 rifle for a half-hour of individual manual-of-arms drills to keep my arm in good shape for the next day, too.

Even now, I hope Homer's arm killed him for the next 36 hours.

Some Soldiers Need to Remember, In Spite of Ourselves

You really do not have to like a soldier to want his name or his person to be remembered. Picture yourself being in the spot where this story starts and ends, at the Army Ranger bayonet assault course, Ft. Benning, Georgia.

No, you can't picture that, and I wouldn't wish that ugly image on anyone, not even the staff sergeant I truly hated that day while running the course. He was in charge of station #4. In plain words, the bayonet assault course was an irregular looped path through a forest, about a half-mile or so in circuit. The Ranger student's job was simple: double-time the circuit and jump into the pretend enemy foxhole pits at various stations and thrust your bayonet into the straw dummy enemy soldier slumped against the sandbags, while yelling Ranger-like sounds. The pits were loaded up with eight inches of water, mud, and everything in between. The number of circuits depended on what the school commanded for any particular class.

The school management must have liked our class because we only had to run three circuits. Most Ranger classes had to do five, so I heard later. Nevertheless, after heading down the backstretch toward station #4, second circuit lap, my 22-year-old voice gave out. Imagine the pretend laughter of the pretend stuffed enemy soldier as this voiceless Ranger hopped into his pit and made a voiceless thrust at his straw-filled chest.

All the fury of an E6 Army Ranger staff sergeant supervising station #4 avalanched my way, in describing the stupidity of my attack. He didn't actually use the "pussy" word I don't think, but came close, along with long strings of ugly nouns. Then followed by his command to hit the "ground" for 20 pushups, mud and all, and then do a headstand in that same slop. Then being told to get out of *his* pit and get out of the Army if I couldn't act like a soldier. Doubtful if this man could make sergeant first class E7, but he truly was an excellent staff sergeant E6.

Circuit #3 re-birthed not only my voice but also had me running at gallop speed bearing down on station #4. Into that pit I gorilla-pounced, howling with two "long thrust bayonet" moves at the dummy lounging and chuckling against the sandbags, then two muddy high-kicks at the shirted haystack, then a leap *backward* out of the pit, before vaulting back in again with two more full-fury jabs into that straw chest. Over my shoulder, I heard that sergeant shouting excited words and almost clapping (Ranger talk for "That's m'boy!"). Little did this staff sergeant know that I had just mentally put four muscular bayonet jabs into *his* shirt. That was the only second-degree murder I have ever fully visualized, before or ever. Right through his starched fatigue shirt, *the goddamned foul-mouthed bullying sonofabitch walking prick bastard NCO.*

It was at this sparkling instant that I "understood" being an infantryman. It first meant that you fought to win, totally. Nothing else would make any sense. General Fred Franks (Gulf War I) said that when fighting an enemy, you want the odds to be 100 to zero, in your favor. Same thing.

The second lesson? My personal job was to bring this personal destruction in a directed way—deliberate action worked with "the plan"—because of, and for my team.

I am sure that the hero of this story has passed along to the Big Orderly Room in the Sky. His name cannot fade from my memory, because I never took time to see it, much less learn it. I shall just name him *Staff Sergeant Bayonet*. You can be sure that his face and his crucial instruction, though, never left me as long as I wore crossed rifles on my collar, thank goodness. Thank goodness! And now I hand this man's memory gently to you. With good fortune some of you may recall and retell this story to someone else long after I have joined this staff sergeant, *and real man*, in the Big Orderly Room. Even if you don't, he and I will still be shooting some pool, or fighting over a hand of Army pinocle.

RVWG
VERITAS PER VIRTUTEM
MMXIV

Dave Cole

1968 - 1969 served with 1st and 3rd Marine Division as infantryman and Squad Leader in both Divisions.

Village of Wayland Police Officer, NY 1975-1977

New York State Police 1977 - 1999

Steuben County NY Undersheriff 1999-2012

Steuben County NY Sheriff 2013 - 2017

Culture Shock

By Dave Cole

I attended St Mary's College in Erie, Pennsylvania, from late August 1963 to early June 1964. It is a Catholic seminary, and I was training to work as a Redemptorist missionary overseas. The College is now called Mercyhurst. Instead of sleeping in a bedroom, we slept in a large dormitory with a priest constantly walking through at night to monitor students to ensure we were sleeping.

A new sport I was introduced to was Canadian handball. It consisted of a large outside court with a wall and two sides. You played with five other seminarians as two three-man teams. Each team took turns returning the small hard ball to the wall by slapping it with their hand or, on its second bounce kicking it back to the wall.

We played a lot of ping pong also. The Puerto Rican students proved to be very proficient at ping pong.

At Easter time, we received large packages of candies and fruits from home as our Easter reward. There would be large poker games using these Easter gifts as your "chips" to gamble for entertainment.

It was a life much different than I was used to. During the year I attended, John F. Kennedy was assassinated. I recall hearing the radio broadcast from the Seminary Office as I sat in Algebra Class.

On one occasion in Algebra class Jorge Figueroa, a fellow student, was chastised by an Algebra instructor (all instructors were Redemptorist Priests) for something he did wrong. Jorge Figueroa remarked back something in Spanish, which probably was not nice, and the English-speaking Priest responded to Jorge in Spanish. I noticed Jorge's face was red, and I realized Jorge didn't know the Priest also spoke and understood Spanish.

Our meals were in a large cafeteria, and a student in a higher grade read the story of Knute Rockne while we ate. At every meal, grape jelly made from vineyard grapes, applesauce from the orchard's apples, and plenty of homemade bread were served. I ate a lot of each at every meal.

The amazing thing about St Mary Seminary was that it had four years of high school classes and four years of college classes. A separation policy existed, and the College Seminarians were forbidden to speak to the High School Seminarians and vice versa.

We attended Catholic Mass every day, which included singing lots of hymns.

One of the funniest things happened when I went to weekly confession. I always used the same Priest and revealed some things I would be embarrassed to mention publicly to him. In the confessional, there was a screen between the Priest and me, so I felt comfortable confessing because he couldn't see me.

One day when I was walking down a hall passing this same Confessional Priest, he stopped me and said, "David, I will be away next confessional day, so you'll have to use another Priest that day." I was so embarrassed that he knew who I was.

Another Wayland boy I knew attended St Mary's Seminary simultaneously. He was much more outspoken and a tinge more disrespectful at times toward the Priests. The Seminary conducted occasional Proficiency and Conduct grades in an open forum with all the students present. I remember this Wayland student received a 65% for both his Proficiency and Conduct from the Seminary. He soon left and returned to Wayland.

My time at the Seminary was good in terms of education and maturity, and it was awful in the homesickness caused by my being away from my hometown of Wayland, New York.

One of my embarrassing points was that I suffered from Osgood Schlatter disease or extreme knee pain while I kneeled, preventing me from kneeling during Catholic Mass. Instead, I was forced to sit while other Seminarians knelt during Mass, and it bothered me.

Being still youthful, I did not return to St Mary's Seminary after the summer break of 1964. A few years later, I found myself, along with other Marines, killing men in Vietnam.

Extremes

by Dave Cole

Jungle rot is an affliction of the skin's surface and is caused by being continuously dirty, a condition often suffered by the infantry in the Vietnam War. During military operations, we generally spent up to a month in the jungle or" bush" wearing the same uniform day after day. We also wore the same underwear and socks. There were no showers to cleanse the body. During a patrol, if we were lucky, we might have the opportunity to bathe in a creek with no soap. We were filthy most of the time, with sores on our faces and arms from exposure to injury caused by cuts from elephant grass or bamboo punctures. It was common to see small fruit flies flying in and out of our sores. We called them "pussy sores."

Dysentery is an infection and inflammation of the intestines, which causes diarrhea that may contain blood or mucus and is usually accompanied by abdominal pain. It is spread as a result of poor hygiene. For example, if we couldn't wash our hands after defecating in the jungle, anything we touched would be at risk. The infection can also be spread through food or water contaminated with fecal matter. It was customary to dig a small hole in an area to use for a toilet break, as there were no public restrooms in the jungle. There was no means to cleanse the body, especially the hands, after excretions.

Usually, when we were out in the jungle in Vietnam, the only water we had for consumption was supplied by the military from their stocks at the main camps. Individual Marines had to carry their water in heavy canteens and a large amount of other bulky equipment. During operations with enemy contact, helicopters could not resupply us with food and water. Vietnam's climate often involved high heat and humidity, making water vital for survival. During periods of heavy enemy interaction, while we Marines were on patrol, we were often forced to drink water from streams or creeks. The Marine Corps attempted to provide halazone tablets to disinfect the water, but they were often unavailable to us. Drinking unpurified water was risky. Temperatures were over 100 degrees almost daily. The lack of water caused a sort of madness and fury and a depressing feeling that our country was deserting us. On one patrol heading upstream in the jungle, I had my squad fill their canteens and drink stream water. We trudged further up the creek, only to discover several dead NVAs lying in the water, most likely killed some days ago by an air strike.

Leeches were a special treat, too. Crossing a body of water was sometimes necessary, and often when exiting to dry land, we would have several long blood suckers hanging from our bodies, engorged with plenty of our blood. After knocking them off, we usually got an infection where they had broken the skin.

Wood ticks often buried their heads under the skin and, if not extracted properly, would result in an infection.

But mosquitoes were the worst. They were a constant problem. The incessant swarms of mosquitoes biting our skin, especially at night on "hole watch," was an absolute joy. We used to say, "The mosquitoes are so big they pick fights with the birds."

Malaria is caused by mosquito bites and is common in Vietnam. It is a severe and sometimes fatal tropical disease. Symptoms of malaria include fever and chills, a general feeling of discomfort, nausea and vomiting, fatigue, and diarrhea. Generally, a person suffering from malaria ends up with a high fever. I had malaria twice, which produced fatigue and a temperature of 106 at one point. Malaria is often fatal during a third infection.

The first time I caught malaria, we were in the flat lands near Khe Sanh. I remember I took a nap and slept soundly. I felt like I was in a trance as I could hear the Marines near me talking but couldn't wake up. Later we received intelligence that a large NVA force was planning an attack on our position. The Company Commander ordered us to withdraw. I had a high fever. I was queasy and staggered as I struggled over the jungle trails and slopes as we moved away from the enemy attack. Fellow soldiers in my squad carried much of my heavy gear as I lurched onward. A helicopter evacuation was not possible until, at last, we reached a safer area. I was finally evacuated for ten days of treatment at the DaNang Naval Hospital.

Starvation was one more curse that affected us while we were in the jungles. Often, enemy presence prevented resupply by helicopter, and we often went without food for days after exhausting our limited supply of C-rations. I remember in the Ashau Valley in April 1969, we had gone without food of any kind for days. Enemy contact prevented helicopter support. The Company needed volunteers to trudge several miles to a safe area for the resupply helicopter to land. When we arrived at the LZ area, they had brought us only a limited quantity of C-rations, causing anger and disappointment after the long and dangerous trek to the location. The lack of food for days at a time was depressing and weakened us, causing mental hardships and stealing our resolve to continue—but somehow, we did.

The Real Story of the Battle - January 22, 1969

By Dave Cole

During 1968 and 1969, I spent time with both the First and Third Marines Divisions in combat in Vietnam. I was fresh out of high school and felt extremely patriotic and wanted to join the Marines. I enlisted in May of 1968 during the Vietnam war and was headed toward an experience I will never forget. I saw human tragedy from war and went from being a teenager to being a man very quickly.

I met many new friends while in the Marine Corps and in the Vietnam war, but there were two Marines, and one day I have never forgotten. Years later, I traveled to Washington DC, to find their names on the Vietnam Memorial, and have had their Vietnam Veterans Memorial inscriptions on my Office walls over the years.

With the advent of Internet Web Pages, I finally located a website dedicated to Alpha Company, 1st Marine Battalion, 3rd Marine Regiment, 3rd Marine Division. When I checked the names of the Marines killed, I immediately noticed two of the Marines I served with who died in a battle on January 22, 1969, were not listed. They were the Marines whose names I had gotten from the Vietnam Memorial so many years ago. I immediately emailed the webmaster and advised him of the omissions. I received an email back several months later acknowledging the omissions, and the names William St. John and Mark Taylor were added to the list of Marines killed in Vietnam from Alpha 1/3.

In 2003 I located a website dedicated to Mark Taylor. I remember the battle so vividly and remember the circumstances surrounding Mark's death. There was a place for Guests to sign his page on the website. I signed it "I was there, a friend."

Another Marine died that day in the same battle. I had served with this Marine in Boot Camp and with Alpha 1/3 in Vietnam. I carried his body after he was killed heroically charging a bunker that had his platoon pinned down. His name was William St. John. I was never able to find a website dedicated to him.

Several years have passed since I signed Mark Taylor's page, and I had gone on with my life, but never forgotten. I was never able to find anything more about Bill St. John. One day in December of 2007, I was checking my email and, out of the blue, received an email from a person whose name was William St John. When I saw the name, I immediately thought of Bill St John, who also died on January 22, 1969. I thought this couldn't be him as I opened the email. As I read the email,I discovered it was from a William St. John, who stated he had seen my message on Mark Taylor's Webpage and had obtained my email address from that. He stated he was the nephew of William St. John, who had served with the United States Marine Corps and died on January 22, 1969, in a battle with Mark Taylor. I suddenly felt chills go down my spine. The past was revisiting me from the awful days during Operation Taylor Common in the Mountains of An Hoa so long ago. He wanted to know if I was truly there and if I knew his Uncle. I asked him to send me some verification of his identity.

He emailed me a day later and told me Bill St. John had received a Bronze Star for his actions during the battle on January 22, 1969. I had not known that until now

We exchanged several emails, and I felt elated and happy. I felt like I had carried Bill's message of bravery home 39 years later. I felt I had finally come home myself. I also told him that Bill and Marks' inscriptions from the Vietnam Memorial have been on my Office walls over the years and I look at them daily.

I remember it like it was yesterday. It was January 22, 1969, and my Company was at Base Area 112 west of Da Nang. Third Platoon left early in the morning for a patrol in the valleys below to search and destroy any Viet Cong or NVA that may be passing through. The Platoon had 20-some Infantry when they filed out of our perimeter for the routine patrol. PFC Bill S.t John was 2nd man in the column as they trudged down the mountain. I knew him fairly well, as we had been together in some Infantry training in the United States in 1968. I recall we used to pick on him and ask if he were related to Jill St. John, the actress. He also was given the name "John Wayne" for his fearlessness in simulated firefights during training. I can still see him with my other friends as we goofed around during training and were full of life.

On that morning in January, Third Platoon left our mountain perimeter at Base Area 112, west of Da Nang. A short time later, my squad leader told us to saddle up. My platoon, 2nd Platoon, was assigned to board helicopters and provide a perimeter for a downed helicopter that had crashed in the jungle below. We loaded into a Chinook and landed quickly at the base of our mountain. We received small arms fire as we formed a perimeter in a grassy area around the downed helicopter. For a few hours, we engaged an unseen enemy in a tree line in a small firefight.

Toward early evening, our Platoon Commander received word by radio that 3rd Platoon, which was on patrol a few miles from us, had made contact and had taken casualties. Two grunts were killed. The platoon was pinned down about a mile from our position. As darkness settled, a decision was made to send a squad of 20 Marines to assist 3rd Platoon.

Within a short time, we heard AK-47 gunfire in the area 2nd Squad had entered. The squad soon ran back to our perimeter, relaying that one of their members had been shot and killed. It was PFC Mark Taylor, whom I had casually known. He had been ambushed by an NVA soldier as he was changing places with the tail-end Charlie position in the squad (last man). Taylor reportedly was exchanging positions with the last man in the column, who was nervous. I remember we were all upset that the squad had to leave Taylor's body.

During the night, I thought about Taylor and the upcoming day. I was scared.

The next morning it was my fire team's turn to walk point. No one wanted to, and I could see it in their eyes. I didn't feel like a hero and was scared myself. However, I had the strong urge to walk point man, which was the front man in a Marine combat column. I volunteered but knew I had very little experience in combat at that time to walk point. My squad leader acknowledged my request and gave me an old C-ration chesterfield cigarette. I remember thinking, at the time, that this would be my last smoke. I was certain I would be killed.

Soon we left the field and entered the wooded jungle toward the area Taylor had died. I am not sure how far we traveled, but soon I saw a man in a camouflage uniform sprawled on the trail ahead. As I moved closer, I could see it was Taylor and that he was dead. My squad leader ordered me to continue past Taylor's body and stop a short distance ahead on the trail. Two other Marine grunts, Sanchez and Jake Priode, were positioned on either side of me off the trail as we waited while they put Taylor's body in a poncho to transport with us. I was very upset about Taylor and remember having my head down crying.

Suddenly, as I looked up, I saw six to eight men running down the trail toward me, about 50 yards away. At first, I thought they were with 3rd Platoon but noticed they did not have any helmets on. I then realized they were enemy soldiers and opened fire on full automatic with my M-16 as I fell to the ground.

They all fled in different directions. I was chastised by my squad leader for not taking more careful aim with better fire control.

I still felt relieved and excited. I had actually survived and found Mark Taylor so he could go home.

My next assignment was to help carry one of their dead, who was covered completely with a poncho. I remember carrying the left front corner of the poncho with 3 other grunts. I also remember asking who I was carrying, and someone said, St. John. They told me St. John had rushed an enemy bunker after the point man, Pfc William D. Farrell, Le Claire, IA, had been killed, and the platoon was pinned down. St. John was a hero from all the stories I had heard on that day. He was later awarded the Bronze Star.

I was completely stunned by his death. I think everyone feared that if they got to know someone too well, they'd eventually be killed.

It was a hard time for me, but I matured into a seasoned squad leader

I have always mourned St. John's death silently for years because of our contacts with the Marine Corps. Being able to finally carry this story to his namesake has given me some peace. It's like I was meant to do this for Bill.

I will never forget January 22, 1969. I will never forget Mark Taylor or Bill St. John. I walked point man to find Mark and survived, and I later carried Bill so he could go home. I remember them every day when I look at their inscriptions on my office wall and remember the many years ago when we were friends and Marines. God Bless our Veterans.

(Excerpted by permission from St. John's Story, David Cole, to be published)

A Medal of Honor

By Dave Cole

This is the story of a Marine in my squad who received a Medal of Honor posthumously.

I made a mistake in my current writings in my book on Vietnam. I had said while on patrol with my squad that I spotted Viet Cong leaving a village as we were crossing a rice paddy and I dispatched a fire team to kill the two or three VC, who had not seen us yet. They intercepted the VC and killed them. The story goes on to explain a nighttime ambush the platoon Lieutenant had me set up that night near the bodies of the VC near the village. The ambush was successful according to my story with us killing many enemy soldiers.

When my time was up in Vietnam I was airlifted away from my unit by helicopter. When I returned home, I received word by mail that just after my chopper lifted off the Landing Zone, the LZ was mortared by the NVA.

Later, talking to a former Marine I served with in Vietnam, he told me that a Marine in my squad I had screamed at for mistakes, saying he would be killed, was, in fact, killed shortly after I left Vietnam.

I couldn't think for all these years why I told him he would be killed, or what he did to get himself killed.

I was also told that my squad was on point and suffered casualties a week after I left.

I wanted to know how my squad suffered those casualties. I finally contacted John Campbell, a historian who forwarded the story of the 7th Marines' losses on Nov 12, 1969, and who was killed.

I re-contacted the former Marine who lives in Virginia, and he was in Vietnam with my old unit after I left for the end-of-duty. He says the Marines killed on Nov 12, 1969, were members of my squad. He also says that the man I told would be killed was one of them. He said the reason I told this Marine he would be killed was because when I previously stated here that I sent a fire team to intercept VC exiting a village was not true.

My friend, a former Marine, said that the reason I told this Marine he would be killed was because when we spotted the VC exiting the village, this Marine and machine gunner ran after the VC of their own volition and intercepted and killed them. I didn't order a fire team out after the VC. These two Marines left my squad column without my direction and I later told this Marine he was going to get himself killed by such recklessness. Such happened a few weeks later and these two Marines are in the story below.

Info obtained from my former Marine infantryman in my squad, Jake Priode:

Dias and Hamblett were in our platoon. Hamblett came around the fighting holes in a gas mask trick or treating on Halloween night. Dias and Hamblett were always in competition with each other and were very aggressive in combat. Remember the ambush we set up on the rice paddy dikes? That day before the ambush we saw several NVA running through the rice plants. It was Hamblett and Dias who chased after them. Hope this helps.

Almost forgot. Hamblett had "Virginia Boy" written on the front of his helmet and he was the gun squad leader. He died of a wound to the chest on the Medivac chopper heading back to the rear.

Hamblett was the gun squad leader, so he would only be attached to your squad. I believe Dias was in your squad because I remember being right behind him on a patrol. He waded through a stream and suddenly stopped in front of me. He lifted up his foot and had the largest leech I've ever seen attached to the bottom of his boot. It's amazing how those small details stick in my mind, even to this day.

The Story—

"Gone but not Forgotten" 52 years ago today, on Wednesday, November 12, 1969, Delta Company, 1st Battalion, 7th Marine Regiment lost 4 of their own heroes when Lance Corporal Robert Bryant Hamblett, Private First Class Jerome Higgins, Private First Class Ralph Ellis Dias, and Private First Class Carl Henderson were killed in action.

On November 12, 1969, 1st Battalion was on a search and clear operation of Hon Nui Tau, Hill 953 in the Que Son Mountains, a known sanctuary for the North Vietnam Army (NVA) from where they would launch attacks on the valleys below. Moving through dense jungle and steep slopes a Marine platoon ran into the enemy in well-fortified bunkers.

The day erupted with the sound of enemy machine-gun, automatic weapons, and Rocket Propelled Grenades (RPGs) fire directed toward the Marine platoon that was pinned down on the rugged slope. The NVA soldiers were well protected from small arms fire, and as the attacking force, the Marines had to find any nook or cranny on the mountainside for protection as the NVA fire rained down on them.

With the intense fire and shards of metal from NVA mortars and RPGs shredding the foliage, casualties from the heavy fire began to accumulate. A reaction force was sent in to reinforce the pinned-down unit but soon found themselves fighting for every inch of ground to advance towards their beleaguered comrades.

The assault was proving costly, supporting arms were called in, and the Marines fell back to safer positions to enable the artillery and air strikes to do their job on the bunkers and caves they had been confronting. After the artillery and air strikes were concluded the Marines moved through the battlefield but the enemy had slipped away, preparing to encounter the Marines again on their own terms. The day's firefights proved costly for the Marines, six of their comrades were killed in the ground action, two died when a Cobra helicopter flying in support went down and crashed, and many more were wounded.

Lance Corporal Robert Bryant Hamblett was the son of Silas D. Hamblett and Helen E. Hamblett of Forestville, MD. He enlisted in the U.S. Marine Corps on September 6, 1968, in Roanoke, VA. He was a Marine Rifleman, age 19, born March 13, 1950. He arrived in Vietnam on February 16, 1969, and was assigned to Delta Company, 1st Battalion, 7th Marines, 1st Marine Division (Rein), III Marine Amphibious Force, Fleet Marine Force, Pacific. On November 12, 1969, LCpl Hamblett was one of those casualties, he died as a result of enemy rifle fire when his reaction squad joined the battle. LCpl Hamblett had been in Vietnam 270 days. LCpl Hamblett was awarded the Silver Star Medal, the Purple Heart Medal with 1 Gold Star denoting the second award of this medal, the Combat action ribbon, the Navy Unit Commendation, the Navy Meritorious Unit Commendation, the National Defense Service Medal, the Vietnam Service Medal with 4 bronze star devices, the Vietnam Gallantry Cross Unit Citation with palm and frame, the Vietnam Civil Actions Unit

Citation with palm and frame, and the Republic of Vietnam Campaign Medal with date banner. LCpl Hamblett is honored on the Vietnam Memorial at Panel 16W, Line 064. LCpl Hamblett was awarded the Silver Star Medal posthumously "For conspicuous gallantry and intrepidity in action on November 12, 1969".

CITATION: The President of the United States of America takes pride in presenting the Silver Star (Posthumously) to Lance Corporal Robert Bryant Hamblett (MCSN: 2486513), United States Marine Corps, for conspicuous gallantry and intrepidity in action while serving as a Machine Gun Team Leader with Company D, First Battalion, Seventh Marines, First Marine Division in connection with combat operations against the enemy in the Republic of Vietnam.

On 12 November 1969, Lance Corporal Hamblett was a member of a reaction force that was directed to assist a platoon that had been pinned down by fire from elements of a North Vietnamese Army battalion occupying a well-fortified bunker complex in the Que Son Mountain area of Quang Nam Province.

Arriving at the site of the beleaguered unit, the reaction force was also pinned down by an extremely heavy volume of mortar, small-arms, and automatic-weapons fire. Realizing the need for immediate action, Lance Corporal Hamblett seized the initiative and commenced maneuvering his men in an attempt to execute a flanking movement. As he was directing his men across the fire-swept terrain, he pinpointed the enemy machine gun emplacement which was the principal source of hostile fire. Aware that one person would attract less attention than a group; he deployed his team back to friendly lines and commenced a lone assault up a steep slope toward the hostile bunker. Undaunted by the enemy rounds impacting all around him; he determinedly pressed forward but was mortally wounded by sniper fire as he darted across an open area.

> By his courage, valiant initiative, and unwavering devotion to duty, Lance Corporal Hamblett inspired all who observed him and upheld the highest traditions of the Marine Corps and the United States Naval Service. He gallantly gave his life for his country. For the President (Signed) John H. Chafee, Secretary of the Navy.

Private First Class Jerome Higgins was the son of Robert H. and Lou A. Higgins of Springfield, OH. He enlisted in the U.S. Marine Corps on December 3, 1968, in Chicago, IL. PFC Higgins was a Marine Rifleman, age 19, born January 20, 1950. He arrived in Vietnam on May 28, 1969, and was assigned to Delta Company, 1st Battalion, 7th Marines, 1st Marine Division (Rein), III Marine Amphibious Force, Fleet Marine Force, Pacific.

On November 12, 1969, during a search and clear mission in the Que Son Mountains, PFC Higgins was one of those casualties. He died as a result of enemy rifle fire when his reaction squad joined the battle. PFC Higgins had been in Vietnam 169 days. PFC Higgins was awarded the Silver Star Medal, the Purple Heart Medal, the Combat action ribbon, the Navy Unit Commendation, the Navy Meritorious Unit Commendation, the National Defense Service Medal, the Vietnam Service Medal with 3 bronze star devices, the Vietnam Gallantry Cross Unit Citation with palm and frame, the Vietnam Civil Actions Unit Citation with palm and frame, and the Republic of Vietnam Campaign Medal with date banner. PFC Higgins was awarded the Silver Star Medal posthumously "For conspicuous gallantry and intrepidity in action on November 12, 1969". PFC Higgins is honored on the Vietnam Memorial at Panel 16W, Line 062.

> CITATION: The President of the United States of America takes pride in presenting the Silver Star (Posthumously) to Private First Class Jerome Higgins (MCSN: 2457197), United States Marine Corps, for conspicuous gallantry and intrepidity in action while serving as a Rifleman with Company D, First Battalion, Seventh Marines, FIRST Marine Division in connection with combat operations against the enemy in the Republic of Vietnam. On 12 November 1969, Private First Class Higgins was a member of a platoon from Company D which was pinned down by fire from elements of a North Vietnamese Army battalion occupying a well-fortified bunker complex in the Que Son Mountain area of Quang Nam Province.
>
> Reacting instantly to the critical situation, he fearlessly maneuvered to the point of heaviest contacts and fired light antitank assault weapons at two machine-gun

> emplacements, killing four hostile soldiers and silencing the fire from that sector. During the ensuing fierce engagement, Private First Class Higgins observed two wounded Marines fall in an open area that was being raked with intense enemy mortar and automatic weapons fire. Undaunted by the fusillade, he boldly darted into the hazardous area, pulled the casualties to a covered location, and then helped carry them to a landing zone for medical evacuation.
>
> Returning to the battle, he pinpointed a principal source of hostile fire and moving to a dangerously exposed vantage point, ignored the enemy rounds impacting all around him as he destroyed the position with a light antitank assault weapon. Private First Class Higgins was mortally wounded by sniper fire as he attempted to return to his unit's defensive perimeter. By his courage, valiant initiative, and unwavering devotion to duty, Private First Class Higgins was instrumental in the subsequent defeat of the numerically superior North Vietnamese Army force and upheld the highest traditions of the Marine Corps and of the United States Naval Service. He gallantly gave his life in the service of his country. For the President (Signed) John H. Chafee, Secretary of the Navy.

Private First Class Ralph Ellis Dias was the son of Melvin Dias and Anna Mae Dias of Shelocta, PA. PFC Dias was a Marine Rifleman, age 19, born July 15, 1950. He arrived in Vietnam on April 28, 1969, and was assigned to Delta Company, 1st Battalion, 7th Marines, 1st Marine Division (Rein), III Marine Amphibious Force, Fleet Marine Force, Pacific. On November 12, 1969, during a search and clear mission in the Que Son Mountains, PFC Dias was one of those casualties, he died as a result of enemy rifle fire when his reaction squad joined the battle. PFC Dias had been in Vietnam 199 days.

PFC Dias was awarded the MEDAL OF HONOR posthumously, the Purple Heart Medal with 1 Gold Star denoting the second award of this medal, the Combat Action Ribbon, the Navy Unit Commendation, the

Navy Meritorious Unit Commendation, the National Defense Service Medal, the Vietnam Service Medal with 3 bronze star devices, the Vietnamese Military Merit Medal, the Vietnam Gallantry Cross Unit Citation with palm and frame, the Vietnam Civil Actions Unit Citation with palm and frame, and the Republic of Vietnam Campaign Medal with date banner. PFC Dais is honored on the Vietnam Memorial at Panel 16W, Line 064. PFC Dias was awarded the MEDAL OF HONOR posthumously "For conspicuous gallantry and intrepidity in action on November 12, 1969". For conspicuous gallantry and intrepidity PFC Dias was awarded the Medal of Honor for his heroism that night.

The President of the United States in the name of The Congress takes pride in presenting the MEDAL OF HONOR posthumously to PRIVATE FIRST CLASS RALPH E. DIAS, UNITED STATES MARINE CORPS, for service as set forth in the following:

> CITATION: For conspicuous gallantry and intrepidity at the risk of his life above and beyond the call of duty while serving as a Rifleman with Company D, First Battalion, Seventh Marines, First Marine Division in the Republic of Vietnam on November 12, 1969. As a member of a reaction force that was pinned down by enemy fire while assisting a platoon in the same circumstance, Private First Class Dias, observing that both units were sustaining casualties, initiated an aggressive assault against an enemy machine-gun bunker, which was the principal source of hostile fire.
>
> Severely wounded by enemy snipers while charging across the open area, he pulled himself to the shelter of a nearby rock. Braving enemy fire for a second time, Private First Class Dias was again wounded. Unable to walk, he crawled fifteen meters to the protection of a rock located near his objective and, repeatedly exposing himself to intense hostile fire, unsuccessfully threw several hand grenades at the machine gun emplacement. Still determined to destroy the emplacement, Private First Class Dias again moved into the open and was wounded a third time by sniper fire. As he threw a last grenade that

> destroyed the enemy position, he was mortally wounded by another enemy round.
>
> Private First Class Dias' indomitable courage, dynamic initiative, and selfless devotion to duty upheld the highest traditions of the Marine Corps and the United States Naval Service. He gallantly gave his life in service to his country. (Signed) RICHARD M. NIXON

Private First Class Carl Henderson was the son of John Henderson of Philadelphia, PA, and O'Nett Henderson of Canton, MS. He enlisted in the U.S. Marine Corps on April 10, 1969, in Philadelphia, PA. PFC Henderson was a Marine Rifleman, age 19, born July 15, 1950. He arrived in Vietnam on October 16, 1969 and was assigned to Delta Company, 1st Battalion, 7th Marines, 1st Marine Division (Rein), III Marine Amphibious Force, Fleet Marine Force, Pacific.

On November 12, 1969, during a search and clear mission in the Que Son Mountains, PFC Henderson was one of the casualties. He died as a result of enemy rifle fire. PFC Henderson had been in Vietnam for 28 days. PFC Henderson was awarded the Purple Heart Medal, the Combat Action Ribbon, the Navy Unit Commendation, the Navy Meritorious Unit Commendation, the National Defense Service Medal, the Vietnam Service Medal with 2 bronze star devices, the Vietnam Gallantry Cross Unit Citation with palm and frame, the Vietnam Civil Actions Unit Citation with palm and frame, and the Republic of Vietnam Campaign Medal with date banner. PFC Henderson is honored on the Vietnam Memorial at Panel 16W, Line 064.

I cannot close this memorial without recognizing those 2 Marine Aviators who died, providing close air support for their embattled 7th Marines Brothers.

Major Howard Boyd "Hank" Henry, U.S. Marine Corps, was the husband of Jacqueline D. Henry. Maj Henry was a Marine Pilot, age 36, born February 18, 1933. Maj Henry arrived in Vietnam on August 28, 1969, and was assigned to Marine Observation Squadron 2 (VMO-2), Marine Aircraft Group 16, 1st Marine Air Wing, III Marine Amphibious Force, Fleet Marine Force, Pacific. On November 12, 1969, Maj Henry was killed in action while flying an AH-1G Cobra (Tail number 68-15080) from squadron VMO-2 in close air support of the embattled 7th Marines. Maj Henry's helicopter (Papa) was hit by enemy fire and crash-landed while attacking a heavy machine gun dug into the north face of the Que Son Mountains. Lt. Dumas and Lt Deane K. Swickard (Papa 2) were attacking the same gun position. This gun was causing considerable problems for the Marines operating in the immediate area and had several Marines pinned down in a bomb crater. The helicopters' attack was made generally west to east. Numerous gun runs were made without effect.

After a couple of runs, Papa 2's weapons systems malfunctioned. The minigun and 40 mm jammed, and the rockets would not launch due to an intervalometer problem. Papa sent us high and dry to the west. From that position, we watched as Papa made a couple more firing passes. Apparently, the gun was mobile and was possibly being moved in and out of a cave. We were close to the gun. Even in the relatively soundproof cockpit of the Cobra, the gun was clearly audible as it returned our fire. On their last pass, we radioed that they were taking hits, and we could see pieces coming off of the aircraft.

At about that same time, Maj. Henry radioed that he was on fire. We could see what looked like a ball of fire in the exhaust, but no flames were visible from the engine cowling. Lt. Dumas and Lt Swickard turned toward Maj. Henry and Lt. Lofton as they turned northeast. At that time, Lt. Dumas radioed that there was a good emergency landing area at their 10 o'clock. Lt Swickard could not remember if they responded or not. Their crash landing looked good. The Cobra appeared to be generally intact after impact. Papa 2 dove down and made simulated firing passes in an attempt to keep the enemy away and locate their exact positions until we could get an H-46 in to rescue them.

There was no sign of them. Papa 2 looked for them and expected them to come up on guard. Nothing. A radio call went out for help, and Lt. Bartlett arrived with his section of Cobras. As he arrived, the H-46 landed near the burning Cobra. The crew chief found both pilots dead. The main blade had struck both. Maj Henry assisted his weapons officer, 1stLt Joseph Lofton, out of the helicopter, and while they were moving away from the downed craft, both men were hit by the helicopter rotor, killing both instantly.

Maj Henry had been in Vietnam for 77 days. Maj Henry was awarded the Silver Star Medal, the Distinguished Flying Cross, the Bronze Star Medal with the V devise denoting Valor, the Purple Heart Medal, the Air Medal with the bronze 10 device, the Navy Commendation Medal with the V device denoting Valor, the Combat Action Ribbon, the Presidential Unit Commendation, the Navy Unit Commendation, the National Defense Service Medal, the Korean Service Medal with 1 bronze star device, the Vietnam Service Medal with 3 bronze star devices, the Armed Forces Reserve Medal, the Vietnam Gallantry Cross Individual Citation with Palm, the Korean Presidential Unit Citation with Gold Frame, the Vietnam Gallantry Cross Unit Citation with palm and frame, the Vietnam Civil Actions Unit Citation with palm and frame, the Republic of Vietnam Campaign Medal with date banner, the United Nations Service Medal (Korea), and the Republic of Korea War Service Medal.

Maj Henry is honored on the Vietnam Memorial at Panel 16W, Line 064. Maj Henry was posthumously awarded the Distinguished Flying Cross for his actions during the Vietnam War, Specifically, on November 12, 1969, 1st Marine Aircraft Wing GENERAL ORDERS: SYNOPSIS: Major Howard Boyd Henry (MCSN: 0-85097), United States Marine Corps, was awarded the Distinguished Flying Cross for extraordinary achievement while participating in aerial flight in Southeast Asia. Major Henry was awarded The Silver Star Medal for his actions on October 22, 1969.

> CITATION: The President of the United States of America takes pride in presenting the Silver Star (Posthumously) to Major Howard Boyd Henry (MCSN: 0-85097), United States Marine Corps, for conspicuous gallantry and intrepidity in action while serving as a Pilot with Marine Observation Squadron TWO (VMO-2), Marine Aircraft Group SIXTEEN (MAG-16), FIRST Marine Aircraft Wing in connection with combat operations against the enemy in the Republic of Vietnam.
>
> On the night of 22 October 1969, Major Henry launched as Section Leader of a flight of two HA-1G Cobra helicopters

> assigned the emergency extraction of a seven-man reconnaissance patrol that had sustained four casualties. Arriving over the designated area, he immediately established radio contact with the team and was informed that only three members of the team were able to defend their position. Unable to establish the beleaguered patrols exact location, he requested a flare ship and commenced low-altitude passes over the area to search for the embattled men. Undaunted by the constant threat of hostile fire and extremely adverse weather conditions, which severely restricted visibility and the area of maneuverability, Major Henry fearlessly directed his aircraft on repeated runs under the obscuring cloud cover.
>
> Maintaining continuous radio communication with the ground unit, he continually encouraged the Marines and requested guidance by visual sighting of his aircraft's lights or the sound of his engine in an attempt to pinpoint the patrol's position. As a result of his low-altitude circling of their position, he succeeded in preventing the enemy from launching an attack. After over eight hours of maneuvering over the reconnaissance team's position, Major Henry directed the insertion of a reaction force and the extraction of the patrol. By his courage, bold initiative and unswerving devotion to duty in the face of great personal danger, Major Henry was instrumental in saving the lives of several of his fellow Marines and upheld the highest traditions of the Marine Corps and the United States Naval Service.

Maj Henry was awarded the Bronze Star Medal with the V Device signifying Valor earlier in his tour of duty in Vietnam (Citation unavailable). Maj Henry was posthumously awarded the Purple Heart with one Gold Star, signifying his second award. Maj Henry had been awarded the Air Medal with the 10 device.

1stLt Joseph Alan Lofton was the son of Lena Holt Lofton and Joseph M. Lofton. 1stLt Joseph Alan Lofton was a Marine Aviator from Akron, OH.

He was born on October 18, 1945. He arrived in Vietnam on September 1, 1969, and was assigned to Marine Observation Squadron 2 (VMO-2), Marine Aircraft Group 16, 1st Marine Air Wing, III Marine Amphibious Force, Fleet Marine Force, Pacific.

On November 12, 1969, 1stLt Lofton was killed in action while serving as the Weapons Officer on an AH-1G Cobra (Tail number 68-15080) from squadron VMO-2, in close air support of the embattled 7th Marines. His helicopter was hit by enemy fire and crash-landed. He was assisted from the aircraft by his Pilot, Maj Henry. While they were moving away from the downed aircraft, both men were hit by the helicopter rotor, killing both instantly. 1stLt Lofton was posthumously awarded the Air Medal and the Purple Heart.

1stLt Lofton had been in Vietnam for 73 days. 1stLt Lofton was awarded the Distinguished Flying Cross, the Purple Heart Medal, the Air Medal with the bronze 1 devise, the Combat Action Ribbon, the Presidential Unit Commendation, the Navy Unit Commendation, the National Defense Service Medal, the Vietnam Service Medal with 2 bronze star devices, the Vietnam Gallantry Cross Unit Citation with palm and frame, the Vietnam Civil Actions Unit Citation with palm and frame, the Republic of Vietnam Campaign Medal with date banner. 1stLt Lofton is honored on the Vietnam Memorial at Panel 16W, Line 065.

Someone once said, "Death is not final until you are forgotten." While I breathe, all of you will live on, and your sacrifices for our freedom will be remembered. Let me just say, "Thank you and Semper Fidelis" from an old comrade and a grateful nation.

Richard Terboss

Richard resides in East Rochester, New York, with Wheaton Terrier, Chloe. His wife, Dorothy (deceased), and he had three daughters. At last count, he has eight grandchildren and four great grans. Their joint fifty-two-year odyssey led them through the usual early marriage struggles, family happenings (both stressful and joyful), and empty nest adventures. Their retirement hobby was photographing lighthouses and range lights from the Canadian Maritimes through the Great Lakes and from Maine to Virginia. They felt, with the help of a little luck (or was it careful, successful risk-taking), that they were treated fairly during their service time. Read further and see what you think..

Army Tales

By Richard Terboss

I'm in the Army now
I'm not behind a plow
They don't give a switch
If I never get rich
I'm in the Army now...

Do you know that WWI song? It uses humor to portray the soldier's lot, but some of it proved all too true. Here are my Army stories, presented in as direct and honest a way as possible. Each of these snippets is a personal experience, as best I remember them. "At War" is never a pleasant place to be, and I honor those whose daily, or even occasional, experiences caused them to be in danger far beyond that which threatened me. That said, here are "Army Tales."

1 How did I Get Here, and Where did I Come From?

I was born in October 1941, the youngest of three Terboss children. My youth came and went quite normally, as did my teenage years, although my love of baseball dominated the latter up to and including Town Team participation. After finishing high school and working at a few nondescript jobs, I enrolled at the Penn State University campus in Scranton, Pa., pursuing an AAS Degree in Engineering. Xerox Corporation interviewed at my school, and in 1962 they offered me a lab technician job in Product Development in Rochester, NY. A drifting kind of mindset took shape when I began night school courses at the University of Rochester. I was not sure what my major should be. On trips home, a three-hour drive, I started dating Dorothy on and off. She had been a high school classmate. We were both quite shy, but in time, we got engaged. After a year, wedding plans started to form. We began to plan for an October 1966 wedding. The Viet Nam build-up of that year messed things up. My student deferment was no longer honored, and my job at Xerox did not qualify. Enlisting in the Army, where I could pick my job, seemed like a better choice than waiting to be drafted.

Soon, I was on a Greyhound bus from Rochester to the Buffalo Recruiting Center, then onto an overnight train to Penn Station, New York City. A chartered bus whisked me (and others) to my new "Here," which was Fort Dix, New Jersey. It was early September 1966. Fort Dix was a long way from my hometown of Hallstead, Pennsylvania.

2 The Best-laid Schemes of Mice and Men...Oft Go Awry

Our wedding plans were now askew because I would be in Basic Training in October.

Basic Training brought forth another scheme gone wrong. Basic Training was eight weeks long, and I did an excellent job of running, jumping, and keeping up with the other "boys." I was 26 then, an old man to some of those kids. I guess you could say that things went normally except for one experience. To pass the Basic Training regimen, we had to demonstrate skill with the M-1 rifle. There were levels of skill awarded depending on how many "hits" we got on the firing range. The top award was "Expert." With the Vietnam War raging, doing well on the firing range was not my idea of a smart way to kick off my military career. I had a plan. I decided to hit just enough targets to qualify and miss on my other attempts. So there I was, firing from standing, kneeling, and prone positions. Targets would go down when I hit them, dirt would fly when I missed, and sometimes I shot a hole in the air. In hindsight, I went about this all wrong. I hit the minimum score with little trouble, then missed the remaining shots. My total score barely qualified. Heh, heh, I thought, no way they make me a rifleman.

Afterward, as I returned from the firing line, I noticed my scorer approaching with a crooked (euphemism for S.E.) grin on his face. "Congratulations, Pvt. Terboss. I haven't had anyone qualify as 'Expert' all day, so I just made you one."

One thing did go right. I flew home to see my fiancée on the one weekend pass that they allowed us during training. We went into replan mode. Fortunately, it was to be a simple country wedding with just close family and friends. That made it easier to replan. Dorothy and I rescheduled for a mid-December wedding when my class would be on holiday break.

3 Mr. Intelligence Corps, have I got a deal for you...

Next up was US Army Intelligence Corps training at Fort Holabird in Baltimore, Maryland. One of the first things I had to do was inform the Army that I was getting married shortly. Another was to try to get my course of study switched from Order of Battle Specialist to Image Interpreter. In this job, you analyze photos and film taken by airborne cameras for valuable intelligence. This was what I wanted all along, but there were no openings when I enlisted. I hoped they had had a few drop-outs by this time, and there might be an opening by now.

This switch was doubly important to me because I had heard the stories. Anyone with the "Order of Battle" MOS (Military Occupation Specialty) might find themselves in Vietnam at a remote base camp, gathering and consolidating information. Way out there in the target zone—that was not for me. I petitioned the school for reassignment, laying it on thick. I told them they were wasting their time and money making me an Order of Battle Specialist. That for four years, I had been measuring and grading images on film, on paper, and on microfiche, that I had good math skills and would be far more valuable as an Image Interpreter. It worked, and I finished third out of a class of twenty-four.

4 150 Mile Passes Go a Long Way (if ignored)

Once my classes started in November of 1966, I could get passes every weekend. I had Dorothy's 1964 Ford Falcon at the base. She was hitching a ride to work in Binghamton, NY, with the father of one of her friends. Every Friday, I hit the road for Hallstead. The trip took a while because I had two riders who helped with expenses. One got out near Allentown, Pa. The other I took all the way to Binghamton. On Sunday afternoon, we reversed the process. Those passes were good within a 150-mile radius of Fort Holabird. Hallstead, PA, was double that distance, a little over five hours of driving. Add the round trip up to Binghamton to drop off and pick up my traveling buddy, and the trip was nearly six hours. This went on for about six weeks without incident—the things one does for love.

We were married on Saturday, December 10th, 1966. On Monday, the 12th, and a whole 40 hours into our marriage, I drove Dorothy to the

Greyhound Bus station in Baltimore at four in the morning. I put her on the bus to Shamokin Dam in Pennsylvania, halfway to Hallstead. Her father met her there and took her home. I arrived back at the base in time to make morning formation. Later in December, I was able to take leave, and we went to Florida for a belated honeymoon. Throughout all this turmoil, we both kept our composure. At age twenty-six, we both knew that "it is what it is." There was no point in making a fuss.

By the time I left the service, a very patient Dorothy had already moved seven times in three years, with four of those moves being returns to Hallstead to live with her parents. Whenever we had to move, everything we owned fit into Dorothy's little 1964 Ford Falcon, plus a small car carrier on top. We were our own version of the Iditarod Trail Dog Sled Race minus the snow. Fortunately, our landlords or Dorothy's parents furnished every place we stayed.

5 Spying on the Spy

During the winter of 1966-1967, while in training to be a spy on things from the sky, I heard that Mom and Dad were concerned that I was in some sort of trouble with the Army. Someone was asking a lot of questions around Hallstead about me. They talked to my basketball coach, English teacher, and others. Mom said there had even been a car parked for hours at the top of the hill in front of our house, with its occupants just sitting and watching. Hallstead is a small rural Pennsylvania town with about 1,500 people then. If anything strange is going on, word gets around fast. Those "spies" were fortunate that someone didn't shotgun them with a load of rock salt and ask questions later.

Unknown to me, I was being investigated and then chosen for advanced training. I found out after graduation when I was escorted, with five classmates, to the Pentagon. There, a Colonel Hopkins told us we were picked to go to Offutt Air Force Base outside of Omaha, Nebraska, to learn satellite image interpretation at a joint services school. We had the option of accepting or refusing. He also said that school graduates are usually assigned duty in Germany. I accepted.

6 Ah, Omaha! The Fresh Air, the Stockyards, Ahh!

With apologies to you Omahans, here goes. I lived in a two-man dorm room on the base. Dorothy had returned to Hallstead (again) and managed to regain her former job at the Binghamton Credit Bureau. They were glad to have her back, even if it would be only for a short time (twelve weeks). My class would end sometime in June 1967. Anyone wearing an Army uniform was not hassled. I was left entirely alone. They figured you were there for a good reason. In Omaha, for something to do, you could go downtown and look at the tall buildings. That pastime was right up there with inhaling the stockyards. On the other hand, I liked walking down to the runway fence and watching the enormous B-52 Stratofortress bombers and C-135 AWACS (Airborne Warning and Control System) flights of the Strategic Air Command (SAC) come roaring over my head on take-off.

The mission of SAC was to have long-range bombers in the air 24/7 in case there was a missile attack on the USA or if a strike took out our own nearby underground long-range missiles. The AWACS had huge sensitive antennas that could pick up incoming missiles. They could also intercept enemy electronic communications, as they did in Vietnam.

As a part of our program, we made two trips to Washington, D.C. The first was for presentations by each government civilian and military organization that comprised the National Security Agency (NSA). My favorite was the Army Map Service. It looked like a great place to work, and I would find out later how well they did their jobs. The second trip was after graduation. It was to get our assignments. We were told that we were now a part of a cadre of only eighty people—in the five military services and the NSA—with this training. I had so many Top-Secret clearances by now that the exact number was probably Top-Secret too. Talk about getting in on the ground floor! This brings up the question: If I'm so important, how come I'm still just a Private First Class?

7 The Best-laid Plans... III

It is late June 1967. On the second trip to the Pentagon, our little six-man group met again with Colonel Hopkins. Keeping in mind that most graduates of our school went to Germany to keep an eye on the Russians and East Germans, we still knew that nothing was certain. The good Colonel informed us that he did not have time to study our records and backgrounds.

"All six of you are good at what you do, or you wouldn't be here. I'm going to make it simple. Alphabetically by last name, the first three of you will go to Germany. I'm assigning the last three to Fort Bragg in North Carolina."

Guess where "T" fell in this scheme—no, not on a jet plane to Germany, but instead on a bus to Fort Bragg..

My grandmother's name was DuVal. She never married. She could have put her own last name on my father's birth certificate. Instead, she entered her lover's name—Terboss. Go figure. I could have been a DuVal and gone to Germany. As it happened, everything turned out OK. After all, I'm here to tell you about it.

Dorothy and I had a great time in North Carolina. As a member of the US Army's MI Corps, I was assigned by the 218th Military Intelligence Detachment to the 18th Airborne Corps at Fort Bragg. When not working, Dorothy and I took advantage of our free time to take walks, drive around town, inhale the magnolias' scent, and see the Dogwoods in blossom—what a change from Omaha! We also took in the beaches near Wilmington, NC. Like all service couples, we managed to have a good time sans good-time money.

8 Bond, James Bond

My workplace at Fort Bragg was in the basement of the Headquarters building of the 18th Airborne Corps. You had to pass through two cipher locks to gain entrance. Real James Bond stuff, that (or should I say "Maxwell Smart stuff"? How did it happen that despite my incredible knowledge and important stature and living off-base, I still had to draw KP duty and occasional barracks night watch? Just ask Dorothy about the fun she had dropping me off at the base for a 4 AM KP stint, then returning home to try to get some more sleep before she went to her job in Fayetteville. Yep, we had fun in North Carolina. A happy outcome of my assignment at Fort Bragg was my receiving two promotions (to Specialist Fifth Class) in less than ten months.

My job was updating or creating contingency plans for getting troops of the 82nd Airborne into and out of hot spots in their sphere of responsibility. That sphere included Africa, the Mid-East, Asia, South America, Central America, and the Caribbean. I found this all fascinating. When not studying films and photos, I spent time cataloging innumerable reels sent from Washington, D.C.

On occasion, I found myself briefing Colonels and Generals. Our unit commander was a reserve Artillery Major who had been recalled to active duty. He admitted that he knew little about intelligence matters. He left the briefings up to us specialists.

9 Students, Take Out Your Geography Books

I always liked geography in school. In 5th grade, I proudly memorized the location of all the states and the names of the capitals. All the textbooks in the world could not match the geography and history lessons that I would learn just by going through reels and reels of satellite and U-2 photography.

The U-2 "Dragon Lady," developed in secret by Chief Engineer Kelly Johnson's Lockheed "Skunk Works," is an ultra-high-altitude spy plane, almost glider-like, with a huge wingspan that flew over places like Cuba and the USSR. It was virtually undetectable until the Soviet Union, in 1960, developed a missile that could reach a high enough altitude to shoot one down in the famous incident involving CIA pilot Francis Gary Powers.

From those U-2 photographs, I learned that half of Cuba was covered with baseball diamonds. They were around every corner and everywhere else as well.

I watched the Egyptians, with Russia's help, build their massive High Aswan Dam across the Nile, almost block by block. Tel Aviv, Israel, almost overnight, went from a beach vacation resort to a thriving commercial and financial center. South America's major cities looked overcrowded and really did have "shantytowns." Africa looked beautiful from the air, but for some countries, it was a time of gaining independence (sometimes peacefully, often by violence) from their colonialist "mother countries."

10 Terboss, The Straight-Leg

Working for the HQ 18th Airborne meant everyone left me alone to do my job. They didn't know what I did in that dungeon in the basement of the HQ building, but they knew it was "very important." They left me alone, even though I was probably the only soldier on the base who refused to go to "jump school" and was not wearing the "Airborne" patch on my arm. I was a "straight-leg." That is a term meaning one who is not jump-qualified. Fifty-five dollars a month of additional pay was not enough to make me want to jump out of an airplane. It turned out that this was one of the best decisions I had ever made.

We were sad that we were at Fort Bragg for only one year, but I got my orders for Vietnam. When I shipped out, Dorothy went back to live with her parents, only this time, she was pregnant. She did have the support of both of our families and our friends. Our daughter (spelled Sharen on purpose) was born seven months later. We managed to communicate by mail and by an occasional taped message.

When my two buddies and I got our Vietnam orders, my two comrades, who were jump-qualified, went to the central highlands near Pleiku, to the 101st Airborne, much closer to North Vietnam and the hot action. I went south, near Saigon, with the 1st Infantry Division.

Come to think of it, I was only 30 miles from a section of the border with Cambodia called "the Parrot's Beak." It was one of the major infiltration routes into Vietnam. The 1st Division's northernmost base camp, An Loc, was right on the border. I was never assigned to An Loc. Still, it was better to be closer to Saigon than up there near the North Vietnam border.

11 I Arrive in Vietnam. Hurry up and Wait

I landed in Vietnam on May 1, 1968, shortly after the major Tet offensive—launched by the North Vietnamese and Viet Cong at the start of the Lunar New Year—was repulsed by US and South Vietnamese forces. The offensive had a three-step goal: First, to overrun Saigon, which was impossible, given the defensive forces there. Second, to discourage the South Vietnamese and make them think that a take-over by the North was inevitable (which turned out to be accurate, but it took another eight years), and last, to discourage America, which had begun to sour on the war.

I had arrived at Long Binh, a supply depot near Saigon (renamed Ho Chi Min City after the North Vietnamese won the war), and awaited my assignment to the 1st Infantry Division for a good two weeks. Others had moved on after a few days, and I couldn't figure out the problem, although I was perfectly willing to wait. As it turned out, I was to be part of an experimental unit called a Combat Intelligence Battalion, or CIB. Division engineers were putting up the HQ building, mess hall, and "residential" framed tents that would be our home in Lai Khe, 20 miles north. It would be large enough to be called a Battalion rather than a Detachment.

Many Army divisions had to share intelligence resources and the gathering of information. The 1st Division already had its own Mohawk reconnaissance aircraft equipped with conventional and infrared cameras. It was now going to get every skill set and technology needed to conduct airborne and ground-based intelligence gathering. The only technology it would not get was Side-Looking Airborne Radar or SLAR. Those images were created by a unit out of Vung Tau, south of Saigon. This service was "hired" by the 1st and 9th Divisions, and they sent images to both simultaneously. There was no need for satellite photos, so my clearance for them was removed.

12 Head 'em Up, Move 'em Out

Finally, I boarded an early-morning supply convoy heading north to the three large 1st Division brigade-level base camps and the battalion-level base camp at Phu Loi. I and my M-16 rifle rode hunkered down below the sideboards in the back of an open truck. The paved road was patrolled and swept for mines every dawn. My big concern was snipers. None materialized.

Along the way, I was amazed at the destruction. Brick buildings, of which there were only a few, were destroyed, with only parts of a wall or two standing. The foliage had been stripped and cleared for a good 70 yards on both sides of the road. Wooden buildings, and even shacks, had been gutted. I knew there were countries where the major modes of transportation were bikes, scooters, ox carts, and "one foot in front of the other," but I was seeing it first-hand. Huts were barely four walls with tin roofs, some with windows, some with only shades.

13 "Well, Here We Are in Beautiful Downtown Lai Khe."

I can just hear Bob Hope saying that at the beginning of his USO show for the troops. I never saw him or his show, but I was told that my workplace in Lai Khe looked out over the field that held his audience when he was here a couple of years ago. The office was a combination large storage container/bunker, half buried in the ground and surrounded and covered by sandbags. Lai Khe was home to one of the 1st Division's three brigades. Depending on its mission, a brigade held 1,500 to 3,000 men at that time. It occupied what once was a working Michelin rubber plantation. There were ex-workers nearby. Their village was fenced off from the central part of the base but was inside its outer perimeter. Some of them entered the base every day to do a job (laundry, general clean-up, cut hair, sell stuff). Others were farmers. We were housed in the middle of a dense growth of young rubber trees, so thick they had almost no root systems. They fell easily in the monsoon storms. This was not a good thing for the framed-tent quarters, but I wasn't aware of any tents being taken out by them. If the Army took down a tree, it had to pay Michelin. If God took down a tree, I suspect they sent Him a bill too. The dirt roads were smooth in the dry season but rutted, bumpy, and puddly-potholed during the wet. This was the first time I had experienced hot rain.

14 Life on the Edge of Town

While waiting for the CIB to get its act together, we had to pull perimeter guard duty. Let me set the scene: We had to man five perimeter bunkers, some directly between the village and "out there." "Out there" was pretty barren for about a hundred yards. Nevertheless, one had to be alert (both front and back) to the possibility of someone trying to sneak in or out of the village. I was Sergeant of the Guard that night (what luck!)

and in bunker three. I was equipped with my M-16 rifle, a radio, a walkie-talkie, and a pair of night-sight binoculars. These were able to magnify light so you could see in the dark. You saw dark shapes on an olive-green background, and you could distinguish and identify the forms.

It was my job to make sure the troops had what they needed and that they were on duty and awake. I could talk with bunkers one and five but couldn't see them. We did not have enough walkie-talkies to supply bunkers two and four, but I could see them by eye and with the binoculars. As I made my rounds, I could see one and five from two and four. The two-man bunkers rotated every two hours. We would be on from dusk until dawn the next morning. Every odd hour, I took a walking tour of the bunkers along a dirt path. Sometimes I covered the entire length; sometimes, I just played the binoculars on one and five. They were probably eighty yards apart. I grabbed a bit of sleep in between, making sure the troop on watch woke me up. In front was the flat "out there," and behind was the mostly friendly village. You had to climb a small hill to get to the base or to the village.

Around dusk, all was well. Suddenly, to my right came BLAM, BLAM, BLAM. it was coming from the direction of bunker five, the one I could not see. The officer of the guard was on the horn immediately, wondering what was going on. I contacted bunker five and was told this story:

"Everything is OK! It's just one of our troops sighting in his rifle."

I got back on the horn to the officer of the guard and assured him there was no problem and that I would take care of it. Off I went at a trot. The troop on the spot was someone I had seen around the CIB, but I didn't know him personally. He was very young and had a New York City accent and attitude. He said he hadn't had a chance to sight in his rifle yet and figured now was as good a time as any.

I informed him in the nicest possible way that:

It was night.

He was near the village where innocent people were trying to sleep.

He was in a combat zone.

He was on guard duty.

The village behind him might have someone trying to sneak out,

Hostile forces were probably right out there across the perimeter clearing, itching for his scalp.

"Sighting in his weapon" under these conditions was an excellent way to get sent home with a tag on his toe.

I could tell the whole time that he thought he was in the right and that I would go away eventually. The rest of the night passed peacefully.

15 Sarge Takes a Ride

In the middle of the night, during a 7 PM to 7 AM shift, we got word that an empty satchel had been found behind one of the infantry company's barracks. The troops assigned to those barracks were away in the field. The bag had contents indicating it had once held explosives, but that was it.

Naturally, MPs and investigators swarmed the area. Nothing was found, but that did not mean that all was secure. Still, we had to take our daily summary and some new photos to G2 Headquarters, about half a mile away. Our three-man shift consisted of a Warrant Officer, an enlisted E-4, and myself. I was now an E-6 Staff Sergeant, and that made me the NCOIC (Non-commissioned Officer in Charge). Usually, the E-4 made the run, but he was only 18 years old—a very polite, quiet, mid-western country boy. He seemed very young and insecure about being in Vietnam in the first place. So I ordered myself to make the run.

It was pitch black, and I had to drive with just the running lights on. I was not authorized to carry a handgun, so I propped my rifle across my left elbow with my right hand on the trigger. I had a full clip loaded. I figured that if a hand reached in from the side, whatever was attached to it would be wearing 16 bullets inside of a heartbeat.

Since both my hands were occupied, I could not shift gears. I would make the trip slowly anyway because I could not see that far ahead with only the running lights for illumination. I steered using short quick, release grips with my left hand. I had to make this trip twice during the night, keeping an eye out for the enemy and praying I didn't run over any explosives. To my great relief, both trips were completed without incident. Then dawn came, and they called off the search and declared an "all clear."

My promotion to E-6 is another little story. The CIB was an experimental unit with slots to fill. I was 27 then but acted older. Then there was my special training and, to top it off, of course, my good looks. I appeared before the Promotion Board without a clue about what they might ask or what "stock" answers they expected. Nevertheless, after only 22 months in the Army, I was promoted to Staff Sergeant. It usually takes a person six or more years to make it. Go figure! Must have been the good looks.

16 To the Bunker and Back

The Viet Cong had enough ordinance left over from the Tet offensive that they were able to fire a mortar round or rocket into our camp once a day, every day. It would occur in the morning, around noon, or at suppertime. We would hear the explosion, a siren, and we would head for the nearest bunker, just in case there was going to be more than one. We came back out on the "all clear," which meant the gunship had found nothing in its sweep. In my four months at the camp, there was never a second round. The CIB was located on the far side of the camp, on the other side of the airstrip, far away from "downtown Lai Khe. "Downtown" consisted of the commander's headquarters in the basement of the plantation overseer's house, the brigade mess hall, the post exchange, infantry company barracks and barracks for their support units, and a chapel. Just about everything was there except the motor pool, which was at the airstrip—my story about the chapel in a minute.

The airstrip was not dirt, grass, or paved; It was PSP (pierced metal planking)—metal sheeting attached side-by-side and end-to-end. Mostly, only helicopters landed there and, once in a while, a small fixed-wing scout plane.

The Viet Cong tried to catch people in groups outside or inside, i.e., at chow times unloading convoys or in their barracks. We could set our watches by it. Before our mess hall was completed, we ate downtown—those waiting to get in to eat stayed under the building in a large bunker-type room. We ate at ground level and scarfed down our food quickly, although the hall was well-fortified.

In my time there, the only round our area received landed about 150 yards away, near the camp's perimeter. No one was hurt, but some rubber trees died or were wounded. We figured some VC mortarman had his coordinates all wrong or his map upside down because it was obvious that we were not the target. We reached a point that if we heard a round land downtown, we stayed put, waiting and listening for the second one—which never came.

17 The Little Chapel in the Woods

During some time off, I decided to go to mass at the little chapel over in "downtown." It was a small building, about one-third the size of our St. Jerome church back home. A large tree was located next to the left rear corner of the building. I went in and sat in the left rear corner for no reason other than that was where my family always sat. After mass, I decided to walk back to my tent, close to a mile away but ended up hitching a ride when I got as far as the airstrip. I had no sooner settled in for a quiet afternoon when there it was again—a boom, followed by the siren. I found out later that a mortar round had glanced off the tree by the chapel and had blown the roof off the section where I had been sitting. If I had attended mass an hour later.... Some things just don't have your name on them.

18 Shave and a Hair Cut, Snap, Snap

Vietnamese barbers would cut my hair at each camp. They would do something extra when they finished and were offended if I did not let them do it. They put one forearm on the top of my head and gripped the side of my face. The other hand gripped under my jaw. Then they rotated and snapped my head quickly to one side. Next, the process was reversed, snapping in the other direction. My neck did make a snapping sound, but I can't say that I felt any better or worse. I was just glad they didn't mess up.

19 Fasten Your Seatbelt if You Have One

Given the sweaty, humid days, I wanted to ask a doctor about some scars on my chest that I have had my whole life. My brother had a heck of a time convincing the Veteran's Administration that his personality disorder was traceable to his service time in Saudi Arabia. I wanted a check-up on record in case the scars got worse. The Division's medical center was at Long Binh, the supply depot outside Saigon. I hitched a chopper ride but, stupidly, forgot to sign out. The doctor I saw told me to keep using the steroid cream I had been using for some time. That was it. Because he never prescribed anything, I seriously doubt any record was made.

By now, it was late afternoon. It had just come to me that I had forgotten to sign out. I had to return to Lai Khe to make my night shift, or I would technically be AWOL. Luckily, the last chopper out that day was scheduled to stop at Lai Khe to drop off mailbags. But there was a problem: it was loaded with passengers. The only seat was the open-air gunner's bench which hung out the side. The gun mount had been removed, and there was no seatbelt. The only handhold was a long rod that attached the bench to the chopper. There was only a narrow footrest with nothing to brace my feet against.

But the weather was good, and it should only be a ten-minute ride. I got a good two-hand, white-knuckle grip on the rod and was thankfully still on board when the chopper landed at Lai Khe.

20 The Magic Coordinates

When I was assigned to the Phu Loi camp, I finally got to work with images on film. I was in the last four months of my tour of duty. In the wee hours of the morning, I would scan reels of infrared film for white dots (fires, clusters of people) and SLAR (or radar) images for black dots (boats or rafts on the Mekong River). The background of each type of scan would show enough shadowy detail of known bodies of water, forest edges, trails, and clearings to let me assign map coordinates to the dots.

During the day shift, we would paste together mosaics of camera photos for the commanders interested in certain areas. These coordinates were called in to the Division artillery units who would fire rounds at that spot. The trajectory of the shells was well under the altitude of the camera aircraft, or they would wait until the aircraft had cleared the area.

One time, Division headquarters called, asking who made the readings on the last run. That would be me. It turns out that one of the hot spots I had plotted was a mortar round going off under the plane. The coordinates I had assigned matched exactly with the radar trace of the same round. They wanted to know who down there could plot hot spots as good as radar. Of course, it was a fluke. There was a lot of space out there, and our maps were not 100% accurate, although I was amazed at how accurate they could be, given the dearth of landmarks. Anyway, our Warrant Officer in charge of the unit was not going to let reality stand in the way of talking up his crew.

21 Fool on the Hill

There was a popular song back then called "Fool on the Hill" by Sergio Mendes & Brazil 66. I liked the song and pretty much everything the group recorded. This song has a special meaning for me. At Phu Loi, we listened to Armed Forces Radio while working. There was a program change at 4:00 each morning. The show coming on had "Fool on the Hill" as its theme song. That meant, as soon as it came on, we could expect the pilots of our Mohawk reconnaissance aircraft to be coming through the door for their infrared recon flight plan briefings, which we had just received from Division G-2. Soon afterward, we would hear them taxi out to the runway, throttle up, and take off. At first, we had to wait for the pilots to land, download their film to the darkroom next door, and then we would get the reels put on our light tables. Just before I returned to the states, pilots began transmitting directly from the air to the darkroom.

22 If You Are Upside Down, It Must Be Australia

All troops were entitled to one week of R & R (rest and recreation) during their one-year tour of duty. My request for R & R finally came through during my ninth month in Vietnam. Hawaii and Australia were the most popular destinations, so it took a long time for your name to work its way to the top of the list. Dorothy and I had already discussed meeting in Hawaii, but it was just not practical. It would have also been costly. And now she had an infant to take care of as well. I only had three months to go before I got back, and then I would be in line for an "early out" discharge, so we decided to tough it out. You could get an "early out" if you had less than six months of enlistment time left when you returned from Vietnam. That was me, all right. I headed for Australia.

I got to Sydney, Australia, in their winter season, that is, if you call 70 degrees F "wintertime." I stayed in the city, taking visitor tours and spending a lot of time just hanging out in the beautiful park by the hotel. I went to the aquarium and the zoo and walked along the beach. The tours to the quasi-Outback would have taken up a couple of days, and I was not interested in things like sheep shearing anyway. If they had had tours to the real Outback, I might have taken one. I bought some furry things for Sharen and shipped them, plus called Dorothy a couple of times. I also slept a lot.

23 Deep in Sleep, Very Deep

At Pho Loi, buried right alongside our shop, were huge metal Air Force shipping containers. They were so tall you could walk upright through them. They were mostly underground, with two feet of sandbags placed on top. They were wired for lights and to power the fans and air conditioners that had been set up inside. Six or eight people could sleep there in the bunk beds. We could walk back and forth through the tunnel connectors they had placed between containers. Entrance-way stairs were reinforced with board walls and had sandbag roofs. It was much like entering a mine.

When my time left "in-country" got down to three weeks, that is where I slept, rather than in the wooden barracks up top. There were absolutely no signs of imminent attack, and I would be the first to know. But I was not tempting fate. Man, it was cold down there. I don't know how trolls do it.

24 We Pick Our Garden Spot

Dorothy and I had discussed what would happen once my Army service ended in the spring of 1969. I was on military leave of absence from Xerox, which meant that for almost three years, Xerox had been putting money into my profit-sharing account and maintaining my other benefits, such as vacation time. I had less than two years to go (nights) for my BS degree, which Xerox, and now Uncle Sam as well, would pay for under their tuition aid plan and my GI Bill benefits.

I felt that I could have gotten a good job at any of the National Security Agencies, except at the Army Map Service (my favorite). Those jobs went fast, and you needed lots of civil service seniority. We also decided we did not want to raise a family in D.C. All signs pointed to our coming back to Rochester, NY. It was near our families and seemed like a great place to raise children, plus our kids would have an excellent selection of colleges to choose from if they wanted. So, Rochester became our home.

I had completed my military service and my year in Vietnam, and unlike more than 58,000 less fortunate souls, I had come home safely to my wife and family.

RVWG
VERITAS PER VIRTUTEM
MMXIV

Paul Gardner

Paul Gardner hails from Western New York State. He was born in Sanborn, New York, and raised in North Tonawanda, Jamestown and Henrietta (Rochester). After his discharge from the Army in 1969, Paul came home and then in 1974 moved to Phoenix, Arizona, residing there for five years followed by three years in Oregon. He retired from the automobile sales business after forty years and now enjoys gardening, woodworking, golf and fly fishing. Exploring America is his passion.

Charitable organizations have been a main focus of his life for over forty-five years. Once the Arizona State Director of the Cystic fibrosis Foundation and currently an active member of the HUGS Foundation in Rochester keep him aligned with reality and the opportunity to give back.

APRIL 1968

It's 1957, Jamestown, New York. My friends David Henry, Alan Macintyre, and I are playing Army. I've been ambushed behind Mrs. Reed's garage. I fall to the ground wincing in pain. In a few moments I'll get up and it will be my turn to be victorious. We are three ten-year-old boys acting out a scene from a John Wayne or Audie Murphy movie. But it doesn't work that way here in Vietnam! It's real, it's permanent, and you can lose your turn in the blink of an eye.

≈≈≈≈

Firebase 14, Kon Tum Province, Central Highlands, Republic of South Vietnam. I'm still here, still alive. I looked at my calendar this morning and it's April 4, 1968. I have 106 days remaining before my trip home. The double-digit fidget was a mere 7 days away. This day was a beauty—perfect sunrise, perfect temperature, and no flights on tap. I stopped by Lt. Rich's tent to make sure it was OK to go into the city of Kon Tum. Drew Taylor and I were going to get some new camouflage fatigues made and then head for the Dak Bla River to go swimming. Lt. Rich said to go ahead and to have a good time.

Drew Taylor and I found a mess tent nearby, and after a plate of something that resembled eggs and spam (at least that's what I thought it was) we headed across the makeshift helipad toward downtown Kon Tum. I quickly noticed my helicopter that was absent from its location where I had done the daily pre-flight only a couple of hours ago. As we got closer I saw a Lt. Noname (pilot) getting ready to disembark in another aircraft. I ran over to him and asked him where Lt. Rich and the OH-23 had gone.

"Lt. Rich taxied a Major out to Firebase 14, a few kilometers west of here. When they got ready to leave, the craft sputtered, stalled, and then refused to restart. Magneto failure, I suspect. I'm going to pick up the Major now."

Another soldier climbing aboard is apparently going to stay with Lt. Rich until they either got the OH-23 started or it was hooked and returned to Camp Enari by a CH 47. I told Lt. Noname that it was my aircraft and my pilot, and therefore I should be the one going to FB 14.

The soldier quickly unbuckled, gave me his flak vest, helmet and rifle and seemed to be relieved that I was going and he wasn't.

The sun was clearly at about the 1000 hours in the sky and its warmth was evident as we lifted off. I waved goodbye to Drew, watched him continue his journey to downtown and quickly turned my focus to the direction of the firebase. The no-start problem, although not common, had happened once before, so with that experience under my belt I'm sure we'd be returning shortly.

On the way there two things happened. I was thinking how familiar and beautiful the Central Highlands were and how they resembled the mountain region of my home state of New York. That thought was quickly interrupted when the nameless pilot keyed his microphone and advised me that Firebase 14 was under attack. We best be careful and get out of there as soon as possible. Honestly, that didn't faze me at all. Young and still invincible, what could go wrong? Good advice nonetheless.

A distant barren hilltop rapidly came into view as we sped along at just shy of 80 knots (approximately 90 mph). I saw my OH-23, Lt. Rich and what was no doubt Lt. Rich's taxi fare awaiting our arrival. It seemed a little warmer now, and the sun, although still up there, was partially blocked by the swirling dust of the OH-23's rotor rotating as the collective stick was raised. This allowed us to land softly on this god-forsaken barren hilltop that must have some importance. Otherwise, why would we be there?

I couldn't get out of the 23's seat quick enough as the Major was chomping at the bit to replace me and get the Hell out of there. Lt. Rich was pretty anxious to leave as well, but we first had to figure out why this OH-23 wouldn't start. Lt. Rich, who always seemed so calm, had a sense of urgency in his voice while explaining the symptoms of the no-starting aircraft that should and would get us off this place if operational. I immediately unclipped the electrical box located behind the passenger seat. At that same moment I heard the familiar sound of mortar tubes making that oh so familiar thumping sound as they left their tubes en route to our location.

Lt. Rich and I made a dash to a nearby bunker as explosion after explosion landed in the center of the firebase, creating a series of loud impacts and plenty of shrapnel and debris showering us and our aircraft.

In the safety of the bunker, we waited for the intermission of the attack. Once we felt a pause was in place, we ran back to our OH-23. It appeared our craft was not damaged, so I immediately proceeded with my inspection of the electrical box, looking for the culprit of our extended and unwanted stay at Firebase 14.

I don't know why this even mattered, but I noticed the once warm bright sun had disappeared under a cloud or maybe it was the cloak of dirt and dust from multiple explosions. I could clearly smell the acrid odor of exploded ordnances with which I had been all too familiar in my experiences at Dak To, just a few kilometers north of us.

As if on cue, Lt. Rich and I made a mad dash to our recently visited bunker as the arrival of mortars and now rockets made yet another appearance on this barren and now pock-marked hilltop. This barrage was answered swiftly by Air Force jets, artillery and helicopter gunships.

Thinking this was an opportunity to resume repairs, we scurried to the OH-23 and again noticed, with the exception of a few holes in the fuselage, that we had sustained limited damage. Our peace and quiet didn't last but five minutes, however, and we were off running for safety once again. Between their exploding rounds and our counterattacks it seemed like the world was coming to an end.

I don't recall how many times we repeated this dash for safety, but it was getting pretty old. Now dirty, sweaty, tired, hungry and thirsty, I could only imagine getting out of this nightmare in the making. What's more, I could hear the unmistaken sound of a Quad 50 caliber machine gun and the rattling of small firearms on the north perimeter of the firebase. It was time to get desperate and find a way out of here.

With another pause in the action, I resumed looking for the cause of being stuck here for all of four hours of having this bullseye on our backs. Lt. Rich, now in the OH-23, was awaiting my signal to ignite the magnetos that would fire the motor, turn the rotors and lift us to safety, a meal, drink and a shower. The broken porcelain insulator loomed large as I realized it was the same part that created a no-start on another 23 just a few weeks ago in Dak To.

I found and stole a heavy-gauge wire from the lighting system and prepared a jumper wire that would bypass the insulator and fire the magnetos. Once fired, they would stay fired until we landed back at the makeshift airfield at Kon Tum. As I leaned forward to give Lt. Rich the thumbs up to hit the starter, an explosion at the rear of the 23 pushed me forward with great force. An instant later as I was again moving in the direction of the bunker, the ground behind me exploded, driving me forward at a high rate of speed and driving my face into the ground and filling my mouth with Vietnam soil. I now know what Firebase 14 tastes like. A warm sensation took over as blood turned my dirty green Army uniform pant leg blood red—my blood. I felt a deep burning sensation in my lower body and my eyes felt weak as I sensed shock was setting in. I felt my eyes rolling back in my head, a feeling that I can recall to this day. I asked God, "Please, not today."

Lt. Rich was nowhere to be seen, and I suspected he was in the same boat as I was in. Moments passed as I became numb and unable to move. I looked in the direction of the bunker, and there was Lt. Rich going out of sight as he dipped below ground level to safety. A split second later he reappeared, realizing that I wasn't close behind. Then he looked at me, and I remember precisely his look of shock as he saw me lying face down in the dirt and crimson-stained earth around me. It seemed like only a few seconds later he grabbed me by under the armpits and proceeded to drag me to the bunker. He said repeatedly, "I got you . . . you'll be OK."

Inside the bunker the medics worked to stop the bleeding in my right leg. Fully conscious and coherent now, I watched as they applied pressure to my wounds. I don't recall what the medics looked like or anything they said, but I can see Lt. Rich looking on with a sadness that took over his normally calm demeanor. I felt cold and dirty. I didn't know how bad it was, but I was now feeling tired as whatever they gave me in that syringe was beginning to work.

Something kept catching me out of the corner of my left eye. I freed my hand from my side and touched something. It was a long piece of metal sticking out of my cheek below my eye. I recognized it as part of the aircraft. It was an eight-inch piece of metal sticking out of my face. Really? I grabbed hold of it, pulled it out, held it up, and showed it to the medic who didn't seem to care as they were focused on bigger problems. I pulled another piece from my chin and felt myself getting weak and numb. I looked around and saw them carry another soldier into the bunker. His hand was badly damaged, and he was directing them to his chest area where his blood-stained flak vest was showing yet another wound. The medics turned me over and dressed the wound on my back and buttocks while another medic gave aid to the other young soldier.

I strained to look at the damage done by the enemy's 81 mm mortar, hoping to see my leg intact the same as it was only minutes before. As I'm viewing the medics doing their best to make me comfortable, I couldn't help but hear the explosions of incoming mortar, rockets, and our deadly response with artillery and other deadly ordnances.

A medevac helicopter had been summoned to transport me and another soldier to an aid station (MASH). Only a few short months ago at Dak To I witnessed three medevac helicopters shot down as they were removing the wounded to safety. The NVA soldiers wouldn't shoot the Huey with a big Red Cross painted on its fuselage when it was arriving. Instead, they'd wait until the wounded were aboard, and on the departure they would open fire.

The medic relayed the message from the radioman that the med-vac was near and it was time to go. As the stretcher bearers prepared me for the dash to the helipad and the incoming rescue ship, I could only think about those who met their demise at the point when they thought they were safe from harm and heading home soon. I was scared like never before as I lay there waiting to depart.

I was loaded on quickly and carefully, which afforded me no opportunity to say goodbye to Lt. Rich. The all too familiar sound of the Huey's engine whining and building rpms for takeoff was overshadowed by the thought of bullets tearing through the paper-thin floors and finishing what my adversaries had set out to do in the first place.

"We'll have you home in no time," the crew chief yelled over the noise of the now ascending med-e-vac helicopter. I braced myself, fearing the worst. Firebase 14 was still under attack, but as we quickly flew away my fear subsided. I could feel and hear my heartbeat as I do at this very moment of writing. I caught the eye of the crew chief and asked him where they were taking me. "Dak To," he replied. "Seriously?" I said to myself. One of the most dangerous places on earth and you're taking me there? The crew chief sensed my concern. Then he said, "They'll clean your wounds and send you to the 71st EVAC Hospital in Pleiku. You're going home."

The sun was shining through the glass in the sliding door of the Huey as I felt us descending. The pilot gently landed near the aid station, and the other wounded young soldier and I were whisked away to the aid station. It would be my last visit to Dak To. Dak To was and will forever be a place that harbors so many memories for me.

I remember a nurse, maybe two nurses started cutting my blood-soaked fatigues, socks and boots off me while a doctor and a nurse rolled me over and surveyed and dressed my wounds. A young black soldier lay next to me, and I could see he was missing a leg. I didn't know it then, but we would be side by side again in a holding area later waiting for a ride to the 71st EVAC Hospital in Pleiku. He was wounded by artillery, a victim of friendly fire.

It must be late afternoon, I thought. I know now that I was wounded and medevac'd about 1130 hours. Whatever they gave me made me very sleepy, and I don't remember anything until I woke up in new surroundings sometime in the early evening. The opportunity didn't present itself to thank Lt. Rich for risking his life until forty years later. I had found him through a veteran's locater service run by a Vietnam vet from Corning, New York. When he answered the phone I said, "Is this Rich? This is Mick. I'm so glad I found you." As I spoke, I choked back my tears and felt the lump in my throat getting bigger by the moment.

"I knew it was you and I knew we'd talk someday," he said. I struggled to get the words out as my speech slurred out the words, "Thank you for saving my life."

"You are welcome." Then he broke into laughter. "This is amazing," he said. "I just told the rat story a few days ago. Capt. Broyer never did talk to me again." For the next hour, Lt. Rich and I talked about what we had done over those forty years. Finally, running out of things to say, I thanked him once more and we said goodbye. The weight of finding him and thanking him was finally lifted off my shoulders.

I lingered in a drug-induced slumber until I reached the 71st EVAC Hospital in Pleiku. Darkness was closing in, and I couldn't have been in my bed but for a short spell when a very large nurse from Alabama informed me that she didn't care if I was a General. "In this ward I'm boss and you ain't nothing," she barked. "You'll do as I say and when I say it." I was in no shape to argue. Within a few minutes I was hauled off to an operating room where two surgeons commenced to operate on me. I was given a sedative—a spinal injection—and I felt nothing but heard everything.

One of the doctors said to the other, "This guy's going to have a problem getting through metal detectors."

The masked surgeon looked at me and said, "We are going to leave most of this metal in you. If we take it all out, you won't have much of a leg left. We got the big pieces and what look like shards of aircraft sheet metal out. You're going home, soldier!"

My nurse—I'll call her Ms. Alabama—had food waiting for me as they wheeled me back to my spot. "Rest up today, honey, 'cuz tomorrow I'll have you on your feet. Who's Boss here?" I answered, "You are." The rest of the day was watching and listening to all these boys moaning, groaning and some crying at what wound they had been dealt. Only God knows what dinner was, but it wasn't long after that last bite that an all too familiar sound of a mortar round impacting the ground overshadowed the groans. Ms. Alabama appeared out of nowhere, shouting for us to get out of bed and get on the floor. I tried to move, but I was hooked up to tubes, wires and a sheet tucked in too tight. Ms. Alabama darted to my bed, disconnected my stuff, picked me up, laid me on the floor, grabbed my mattress, covered me with it and said, like she meant it, "Boy, you stay under that mattress until I come get you, you hear me?" Who was I to disobey the Boss? Mortars and rockets were exploding, sirens blaring and our defenses were firing back at a greater rate than what was incoming. Fifteen minutes later all was quiet, no direct hits on our ward, not sure about anyone else. Ms. Alabama, as promised, helped me back into my bed.

Two days later I was being prepared for my next stop at Cam Rahn Bay and then off to a U.S. hospital at Tachikowa Airbase, Japan. As they wheeled me out on a stretcher, I spotted Ms. Alabama. I threw her a kiss. God blessed her and thanked her for everything. As they readied us to load on a C-130, she walked by, tugged my green Army blanket and winked. As I lay there waiting for takeoff, I felt something near my stomach. I lifted the blanket and, low and behold, a Hershey candy bar. Ms. Alabama—what a lady!

A one-night stay in Cam Rahn Bay was all that was needed. I don't recall much of that stay at all. The next morning, we were placed on an Air Force jet and transported to the U.S. military hospital at Tachikowa Air Base in Japan. This place is beautiful in every way—the facility, the care, beautiful nurses, and safe. I would spend three weeks there going through a painful rehab. My Achilles' tendon had been damaged and would cause me unbelievable pain while rehabbing. I was bed-ridden for the first nine days. The food was pretty good, but I heard of this great food at the mess hall and looked forward to going there.

In the bed next to me was a young black Marine from Georgia. Half of his face was heavily bandaged. He was all tubed up, and nurses and doctors attended to him around the clock. He had arrived there a few days before me. About a week later he would become alert and talkative. His wounds were caused by an NVA artillery round causing extensive damage to his right side, but the worst was the loss of his right eye. After a few days of cordial conversation, I got up the nerve and asked him, "How do you feel about the loss of your eye?" A minute or two went by before I received a reply I never expected. He looked at me and pointed to his remaining eye and said, "I'm so glad I still have this one." The one thing all veterans shared—Black or White—was that our blood was red . . . American red.

≈≈≈

Growing up in the little, and I mean little town of Sanborn, New York, had its advantages. First, you know everyone. Second, when the firehall sounded its siren, you knew it was noon. Finally, everyone knew Charlie Treichler. Charlie and I have baby pictures together. We played Little

League baseball and we played Army soldier. Charlie lived next door to my grandparents, and when I would spend summers there, we hung out.

I was on convoy duty in August, 1967. I received a letter from Grandma a couple of days earlier that Charlie was in a place called the Oasis. When we arrived, I realized this firebase was larger than I expected, and the needle in a haystack quandary came to mind. While they were unloading my truck, I thought I'd wander around to check this place out. I come to a dirt cross road. Trust me, this was not any kind of intersection you would know. I stood there for a moment, when as I was about to light up a smoke, someone shouted out my name: "Hey, Paul."

I stopped the smoke lighting procedure, and I looked up at a truck in front of me, and there was Charlie peering down at me. From that point on, anytime I was at the Oasis or Charlie was at Camp Enari we would hang out.

Moving ahead, I'm in the hospital in Japan, 2500 miles from Camp Enari. A Marine from Hawaii was wheel-chairing me to the mess hall. We were rolling along nicely headed for breakfast when all of a sudden my Marine buddy was walking along side me and I was still moving. I turned to see who was pushing me and guess who? Charlie! Walking behind Charlie were a male and a female nurse. I didn't pay much attention at the time, but they were from the mental health section of Tachikowa. I didn't see Charlie for a couple of days until one afternoon while I was having lunch with my Marine buddy, I noticed Charlie at a distance. Seated with him were two orderlies and a young doctor.

The doctor stood up, walked to my table, sat down, and asked me where I was from. "New York," I said. "Why do you ask?"

"What town?" I sensed it had something to do with Charlie, so I said, "Sanborn."

The doctor seemed shocked. "Charlie has been telling us for days that someone he grew up with was here." The doctor went on to tell me that Charlie had been having some issues.

"We never know what to believe as he often slips away from us, if you know what I mean. My God, Paul, would it be OK with you if we gave Charlie permission to visit you and take you around in your wheelchair? It would do him wonders."

"Of course," I said. So, it turned out that every day Charlie would wheel me to breakfast, lunch and dinner, providing he wasn't in what he called "a session."

A couple of weeks later I received word I would be going Stateside. I broke the news to a disappointed Charlie. That night at dinner the young doctor I had talked to came to my table.

"Paul, you'll never know what it meant to Charlie seeing you here." The doctor went on to say that when Charlie first told us about you, we figured it was just another one of his imaginary friends. "Thank you for letting Charlie interact with you. His road to recovery leaped forward tenfold. We'll be sending him home soon."

In a month or more I would get a convalescent leave from the Army hospital in Ft. Devens, Massachusetts. I went to Sanborn after a few days in Rochester to visit my two favorite human beings—Gramps and Gramma. As I stepped out of the car, I looked down the street toward the old fire hall and saw the silhouette of a young man walking toward me on the shoulder of Ward Road. I stood on the driveway and slowly watched as the silhouette slowly and clearly came into view. It was Charlie. A handshake and a chest bump signaled that we both made it home. Charlie's wounds were deeper than mine and would last forever. In that regard, I was the lucky one.

The trip home to the States on a stretcher was what you might expect. I must have slept most of the way because the first stop seemed like a ten-minute flight to Anchorage, Alaska. The Red Cross couldn't do enough to make our journey more comfortable. They took us off the airplane on our stretchers to an area at the airport and fed us breakfast. We left Anchorage a few hours later, and what we were about to experience was unexpected. The aircraft landed at McGuire Air Force Base in New Jersey.

As the pilot taxied to the terminal he came on the intercom. "Gentlemen, in about three minutes I'm going to open the back of this aircraft. Please be ready." The three minutes passed and the aircraft came to a halt. The noise of the hydraulics and mechanical sounds of the ramp lowering revealed a beautiful, sunny and warm day. As our eyes fixed on this beautiful New Jersey morning, the pilot came back on the intercom: "Gentlemen, Welcome Home." After a few moments of dead silence and, as if on cue, the entire aircraft of young soldiers exploded in cheers. "God Bless America" was on everyone's lips, as tears rolled down the cheeks of everyone onboard. As they carried us off the plane, an honor guard on each side of the ramp saluted as we passed.

We arrived at the hospital and were no sooner settled in when the Red Cross nurses made their way into the hospital ward stopping at each bed, now mine.

"What would you like to eat?" one of the nurses asked.

"Whatever you have, as long as it's not in a green can," I said jokingly. Trying to be funny, I said, "I'd like a cheeseburger, fries and a vanilla shake."

"Your mother is waiting by the phone in Rochester," said the nurse. "Would you like to talk with her while I get your food?"

It had been almost a year since I spoke to Mom. I talked to Mom, then Dad and all of sudden I felt home. My fifteen minutes of phone time was up, and as I handed the phone back to the nurse a tray appeared with a cheeseburger, fries and a vanilla shake.

The next day we traveled to a military hospital by air to Fort Devens, Massachusetts. My health and mobility would increase tenfold while there. Another soldier and I took in some great sightseeing in and around Boston. I was eager to go home. In a few days, I would take a convalescent leave and finally go home. I never doubted I'd be home again. I wasn't as healthy as when I departed, but I was going home.

****

"APRIL 1968" appears as Chapter 14, in *Once a Boy*, by Paul Gardner, Mystic Dragon Publishing, 2021

"WELCOME HOME"

A funny thing happened to me when I came home from Vietnam. Nobody asked me where I'd been for the last year, and I never volunteered to tell anyone that I had been there, what I did, or what happened to me.

≈≈≈

You've heard the story before. The American soldier was being held responsible for the war in Vietnam, not the president, not Congress, not the Senate, and not big business that profited. Nope, the American G.I., the twenty-year-old kid from Anytown, USA, was the bad guy. He was ridiculed in public with no welcome home, no parade, no respect, just blame and shame.

My first welcome home was on an airplane coming home from the hospital, but let me give you a little background first. It had been a year since I returned home. I'd spent three months in hospitals in Vietnam, Japan, Maryland and Massachusetts. I finally got the OK to go home on convalescent leave. The Army hospital was located in Fort Devens, Massachusetts, the airport was Boston's Logan International and my ticket was to Rochester (New York) International. Now, Rochester International is kind of a stretch. I think it got the international tag when a Canadian Cessna made an emergency landing in 1956.

I was getting ready to leave for Logan International when my doctor, Capt. Newsome, asked me, "Do you have any civilian clothes?"

"No," I said.

"Well, maybe you should get some," he replied.

"I have my uniform, sir."

"Listen, Paul, Army uniforms can become a target for those who don't necessarily approve of what's going on over there."

"Really? Well, I'm not buying clothes."

"Well, don't say I didn't warn you," he replied while peering over his spectacles.

When I arrived at Logan there were no protesters, but I did get the evil eye from a couple of hippie types. I wasn't worried, but I did feel some kind of relief in that I wasn't being ridiculed or spat upon. I boarded the airplane and took an aisle seat next to a grandfather-aged, white haired man. I'm going to call him Mr. Whipple. Many years later I met a man walking by my house. He would stop and visit every day while I was putting in a new sidewalk. He reminded me of the grandfather on the plane. His name was Mr. Whipple. We struck up a conversation for about ten minutes on our one-hour flight. He saw my cane, and coupled with the uniform he put two and two together. He was dying to ask, so he did.

"I hope you don't mind, but were you wounded in Vietnam?"

"Yes, I was."

"Are you going home?"

"Yes, for the first time, sir."

"Will your folks be there to meet you?"

"I hope so. It was kind of a short notice." He paused for a few minutes and in a lower voice said, "We lost our grandson over there."

"I'm so sorry, sir," I said as I glanced at this nice old man much like my own grandfather.

"He was an Airborne soldier. He died last November in a place called Dak To."

You could have smacked me with a ton of bricks. "Sir," I said, "I am so sorry, I was there. The 173rd Airborne, right?"

His head turned quickly toward me. "Did you know him?" he asked. "Were you with that unit?"

"No sir, but I do know that the 173rd fought there and fought bravely. I'm so sorry." Our communication ended abruptly as I could see sorrow overtake him.

As the overhead 'FASTEN SEATBELT' sign lit up, the stewardess announced that we were on our final approach. I couldn't see my family yet, but I'm sure they were there someplace. As the aircraft came to a stop, the safety-belt light went off, and I unbuckled my belt. I still tear up when I think of what happened next. Mr. Whipple put his hand on my knee and said, "Welcome home, son! I'm happy for you and your family." He extended his hand. I shook it and said, "Thank you, sir."

I grabbed my carry-on and my cane from the overhead and made my way to the exit. I had a newspaper in hand when I saw my mom, dad, sister Kay and Lori following close behind. I opened the paper to conceal my face, and as they drew near I dropped it. My mother let out a shrill as she saw me. We hugged a long time . . . and I mean a long time. Dad, then Kay, then six-year-old sister Lori repeated the process. My brothers Rod and Scott waited at home so as not to tear up in public. Two real softies! Our home was our next stop, and as we started walking to the exit and our ride home, Mr. Whipple, who had been standing in the background, approached us, introduced himself and first shook my father's hand and then embraced my mother.

"I'm so happy your son made it home. God bless you!" Mr. Whipple turned as he walked away and waved at me and saying, "Welcome home." Other than my family and other Vietnam veterans, it would be fifty years before another meaningful "Welcome home" greeting would come along. I'll get to that shortly. Vietnam veterans don't wear T-shirts and hats to broadcast to the world that they are veterans. They wear them to attract other Vietnam vets.

≈≈≈

It's the year 2017 and I'm returning home to Rochester from Tampa, Florida. My baggage was checked and I had about an hour to kill. I was approaching an escalator when off to my right I spotted a man about my age sporting a Vietnam Veteran hat. I made a hard-right-hand turn, and as I got closer I spoke the magic words: "Welcome home, brother." He quickly replied, "Welcome home, brother." We shook hands and had the normal conversation that included what years we served, our unit, what part of Vietnam where we stationed, and our MOS (Military Occupational Specialty). This wasn't my first conversation with a Vietnam veteran by any stretch. We talked for ten minutes or so before I excused myself from our conversation, shook hands, and we hugged each other while repeating, "Welcome home, brother," something all Vietnam veterans say to one another. I walked toward the escalator, and about halfway up I glanced back at my recently discovered 'brother.' His wife had her arms wrapped tightly around her man as he sobbed uncontrollably. I realized at that very moment that we shared something that no one could possibly understand but us. You can't buy, lease, borrow or appear on some silly reality show to experience what we went through while we were over there and years, and even decades after we got home. So when you see a

Vietnam veteran with his hat or T-shirt depicting his service, it's not for you. He's looking for one of his brothers.

The hands down best "Welcome home" would take place one year later. See if you agree. Three friends—Mike Mitchell, Dave Trobia and Mike Judy—and I were golfing in Naples, Florida. We went to an upscale area called Mercato that boasts some pretty high-end restaurants and shops. After our visit to Rocky Patel's Cigar Bar we went to a restaurant close by for dinner. Service was rather poor and our meals came at different times. The two Mike's and I finished and Dave's meal was sent back. We asked Dave if he would mind if we walked around and we'd meet back at the car.

Being a coffee addict, I recalled seeing a coffee shop called the Second Cup on our way to Rocky Patel's Cigar Bar. One of the Mike's went home, and the other went with me. "I need to get a cup of coffee, Mike." So I walked in. There was one person in line, so I stood there patiently awaiting my turn. I couldn't help but notice the man in front of me was Asian. As he turned I reckoned him to be Vietnamese.

"Excuse me," I said, "are you Vietnamese, sir?"

"Yes, I am."

"How are we treating you here in America?" I asked.

"We love it here," he said. He asked me if I had been in his country during the war, and I said that I had been there a long time ago. He thrust his hand toward me and while shaking my hand he said, "Thank you for what you did for us."

"You're welcome. I wish we could have done more," I said apologetically. He went on to tell me that an American soldier had sponsored his mother and father and brought them to America. The Army soldier gave them a place to live, fed them and clothed them until they could take care of themselves.

"We are very grateful to America and its soldiers," he said compassionately. You would have thought I was that soldier whose kindness made such a difference in their lives.

"Well, you are welcome," I said. He continued to the end of the counter to get his beverages, and I remember thinking what a grateful person he must be that he would take the time to thank me for just being there. I ordered my coffee, paid for it and made my way to the end of the counter. I picked up my coffee, turned to find Mike so we could return to the car. But my path was blocked by the Vietnamese man I had just met, although this time he had with him his wife and his twin daughters. They were a beautiful family, nicely dressed, incredibly polite with two well-disciplined young girls probably in their early teens.

He walked toward me and said, "I would like you to meet my family." He went on to say, "I told them what you did for our country, and they would like to thank you and welcome you home." He introduced his wife and two daughters to me, and as I hugged and thanked them each for their kindness, I felt a blanket of warmth come over me. The mother of all "Welcome home" greetings was taking place right in front of me . . . and I knew it. I shook the father's hand once more and thanked him once again for his kindness. I looked around for Mike, spotted him at the door, and started walking toward him. I glanced back at that beautiful family as I got to the door. My ear-to-ear smile must have been a dead giveaway.

Mike couldn't wait to ask. "What was that all about?

"I'll tell you about it on the way home," I replied. This "Welcome Home" was special. In talking to hundreds of Vietnam veterans over the years I found that almost to the man the "Welcome Home" meant more coming from another vet. We shared an experience and we performed what was asked of us, but the treatment we received was like no other in our history. Over 58,000 young men and women made the ultimate sacrifice, and now, decades later, they want to thank us? They had their chance to welcome us home fifty-plus-years ago, but they blew it.

****

"'WELCOME HOME'" appears as Chapter 15, in *Once a Boy*, by Paul Gardner, Mystic Dragon Publishing, 2021

RVWG
VERITAS PER VIRTUTEM
MMXIV

Charles F. Willard

Chuck was raised in a rural setting, tending the family's animal herds from age six. He attended Syracuse University, studying engineering and enrolling in Air Force ROTC. This led to a four-year career as a pilot of C-130 Hercules aircraft, flying missions to 42 countries, including numerous operations in Viet Nam and Southeast Asia. In his subsequent engineering and management career at Xerox, Chuck led the technical team that developed the world's first laser printer. Now retired, he enjoys family, golf, biking, woodturning, and writing. In addition to veterans' stories, he has a trilogy of three unpublished historical novels to his discredit. He deludes himself that he could resume his old pilot's job at any time since sixty-some years later, the USAF is still flying C-130s.

Then (August 1960) and now

RVWG Meeting, 6 April 2022

Old Soldiers Do Die, But They Don't Fade Away

Prompt #1: Soldiers are Special....

By Charles F. Willard

I should amend that to read: *Freedom's Soldiers are Special.* Every day on the news, we see that a totalitarian state's misbegotten soldiers are not "special." What we witness is not "special"; it is a monstrous and malevolent evil. Is the butchery at Bucha what happens when soldiers don't know why they are sent to fight and die? Is it the frustration and rage from not having a higher calling—or any kind of calling—that leads to such mindless barbarism and senseless brutality?

In their yearning to be free and in their devotion to preserving freedom's way of life, Freedom's Soldiers *are* Special. First, they send their loved ones away to safety, kiss their lips and cheeks, and wave goodbye, hands pressing toward each other on opposite sides of the window glass as the train pulls away, fearing in their hearts that they may never see each other again. Then, Freedom's Soldiers go off to fight in defense of their native land and charge into the maelstrom of battle, knowing the cost may be their limbs and their lives, perhaps never knowing whether their sacrifice will preserve their way of life or whether it will all be for nothing.

Gallantry is a word we don't hear much anymore, don't use much anymore. It is a word from the past: The three hundred Spartans at Thermopylae, the battle of Bunker Hill, the Alamo, Meuse-Argonne, Mount Suribachi, Chosin Reservoir, Khe Sanh—bravery against all odds. Now we see present-day examples unfolding before our eyes in Ukraine.

We—each of us, each of Freedom's Soldiers—have a duty and a charge: to keep the special nature of Freedom's Soldiery—and its sacred mission, the defense of freedom—from *ever* fading away. Our life-long obligation is to remind our loved ones and the others around us that Freedom's Soldiers are Special and that the awareness of what they do, what they suffer, and what they stand for cannot be allowed to fade away. If they do, what will we become? Who will be left to fight for our country, our freedom, and our way of life? Freedom's Special Soldiers *will* fade away, and tyrants will trample the face of the Earth.

Writing Prompt for RVWG Meeting, 2 December 2021

Prompt: Write about that one veteran. Think about a veteran, one in particular, about whom you would like to write.

The Service Station

By Charles F. Willard

I have wanted to tell this story for a while. Now is the time. But, as usual, I can't seem to follow directions. This is a story about four people acting together during World War II. How to choose one? At the end of the story, see if you can pick just one.

In 1938 my bachelor uncles, Walter and George, took out a $5,000 loan and purchased a small service station in the tiny hamlet of Alloway, New York. The years before the war were still depression times, and progress in paying off the loan was slow-to-non-existent. They held on, barely. Walter had schooled himself in radio repair, and the extra few dollars from that sideline helped keep the business afloat during those hard times. They were beginning to get their heads above water when the war came.

George, nine years younger than his brother, was 27. He was drafted into the Army, leaving my Uncle Walt alone to run the struggling business. While still a civilian, George had taught himself to type. It was a skill the army needed, and he ended up in a relatively safe clerk job.

Walt was 36 then, not a prime candidate for the draft. But the head of the local draft board, a well-known crafty businessman in town, held an old grudge against Walt's father. Now he had a way to get even, and he made sure Walt's number came up.

Having no choice, Walt went off to war. Unfortunately, his radio knowledge did not work to his advantage. Instead, it "won" him the position of front-line radioman, a well-known short-survival job. It was the radioman who carried heavy spools of cable wherever he went. He ran the vital wires through hostile fire to the most advanced positions, so those taking the battle to the enemy could maintain critical communications with the rear.

When Walt was drafted in 1943, no one was left to run the service station. The little business seemed sure to go into foreclosure. Would the boys' father be saddled with the debt? Perhaps the grudge-holder had his eye on a cheap acquisition.

Not so fast. Marian, the boys' sister, was a registered nurse at Homer Folks Hospital in Oneonta, in the state's southeast region. With the family in crisis, she quit her job and came home to help. Walt's fiancée, Amelia, pitched in too.

Those two ladies ran that little service station, putting in eighteen-hour days, six days a week, for the duration of the war. They pumped gas and cleaned windshields. They changed oil and tires and sold soda pop and sundries, restocking and cleaning up in the off-hours.

When the boys came home safely in late 1945, the girls had paid off the entire $5,000 mortgage. Walt and George now owned the business free and clear. So who are the true veterans of the conflict? Can you pick just one hero from this story? Or are they all "United in Service, United in Sacrifice"?

Flying Formation

By Charles F. Willard

Two T-37 Tweetie-birds sit at the take-off end of the one-hundred-fifty-foot-wide runway, brakes locked. Lead plane on the right half, Number 2 on the left, slightly behind Lead.

Over the radio from Lead: "Power 100%."

"Roger, 100%," from me, Number 2.

Lead's helmeted head gives a sharp nod forward, and we release brakes together. Our take-off roll begins as though we are one aircraft. Quickly, it gathers speed. Lead's job here is to steer a straight, steady course on his side of the runway as we accelerate toward lift-off. My job as Number 2 is to maintain spacing, with only feet separating us. Easier while we are still on the runway, with nosewheel steering, but as the nose wheel lifts off, the spacing must be held by a feather touch on the rudder pedals. No ham-footedness here, or it's a runway smash 'em up. Could be fatal. Now the two of us break ground. Usually, we do this together.

But what if my plane has newer engines or a few more RPMs at the 100% setting? I'll begin to fly before Lead breaks ground. Then things get a little dicey. I now have the wind to deal with. It might push me into Lead if I'm slow to react. And I can't overtake Lead, either. If my engines are a little more souped-up than Lead's, I'll have to throttle back a skosh to stay in position. Risky when my airplane is just beginning to fly. It might plop back onto the runway if I'm not skillful and careful. Messy on a formation take-off.

But we make it off the ground, and now we are both flying, bumping along together at low but gathering speed, clawing for altitude through the morning thermals rising from the sun-heated terrain at the end of the runway. Lead signals for "gear up," then "flaps up." We do these things together so the drag on our ships will recede at the same rate and at the same time. Lead's job now is to pilot his aircraft with the smoothness of a slow waltz and glide it through the air like a swan on still water—quite a challenge with turbulence rocking us around like Jeeps on a washboard road.

Meanwhile, my job is to stay with him, to maintain my position whatever he does. So, my left hand is zipping the throttles up and down in quick little jerks, and my right hand is churning the stick all over the cockpit like a mad cook whipping egg whites, just trying to stay in position through the bumpy air. Then, as we gain altitude, the air settles itself, my motions become a bit less frantic, and Lead tilts us into a 30-degree-bank left turn. As he turns into me, I flick a glance up his left wing and over his canopy at the billowy white clouds hovering just off our course—one of the joys of flying: traipsing amongst the puffy clouds.

The purpose of the turn is to let Numbers 3 and 4, who have taken off right behind us, join up with us. When that maneuver is accomplished, we will spend the next hour or so practicing all the nuances of four-ship formation. The shallowness of our turn allows 3 and 4 to turn inside us. That shortens their intercept path, and when they get near abeam of us, they can slide into position off my left wing. Once joined, we can do all sorts of fun things together.

I risk a quick look over my left shoulder. Here they come, barreling up at us from their inside position. This is a tricky moment. If Number 3, leading the join-up, is too aggressive, he could overshoot and come careening into us, a giant fireball ensuing, so he has to do this just right. On my first formation flight, an orientation with my instructor, we were Lead. Another student, practicing the join-up overshot and came screaming over our heads, missing us by just enough. Scary!

But today we are good. Number 3 has slid smoothly into position on my wing, and we are now a four-ship in left echelon.

We fly like this for a few minutes, doing a few gentle maneuvers. Then Lead gives the signal, and 3 and 4 move a bit to the rear and down, slide under us and join on Lead's right wing, and now we are in a "finger-four," with me on the left and the other two on the right.

At this higher altitude, the air has smoothed, allowing me to tighten my position on Lead, a moment I have been waiting for. A secret "rite of initiation" for formation flying has developed among us students. The T-37 is a twin jet aircraft, the engines sitting close to the plane's body in the wing roots. If the formation pilot can put his wingtip close enough to that jet exhaust, it will come home smudged with carbon.

From my closer position now, it takes just a few seconds. I ease the wingtip in. Sweat is beginning to leak down my face from the tension. Stick and throttle movements are rapid but infinitesimal now. If I screw up, I might be dead. So might the rest of us in this formation. I see the heat radiation of the exhaust flirting with my wingtip, which is now vibrating in the slipstream, only a foot or so from Lead's fuselage. One false move....

Slowly, carefully, I ease my craft away from the peril and return to my normal position. I have come away undetected and unscathed. But I have my smudge.

Looking back, it is a badge of stupidity, not honor, a calculated risk, not only of my own life but the precious lives of the others. One tremor of turbulence, and it could have turned out differently.

Instrument Flying

By Charles F. Willard

Vertigo. My world starts an unsteady wobble, then heaves into a topsy-turvy somersault. It is a sudden nauseating undulation, a lurching, uncertain tumble. My lunch of an hour ago tries to reverse course, pressuring up my swallow pipe and pushing bile into the back of my throat.

I am "under the hood," an Air Force student pilot strapped into the back seat of a Lockheed T-33, the trainer version of the old F-80 fighter plane from World War II and Korea. This is a lesson in instrument flying. I cannot see outside the aircraft. The only light comes from the gauges and dials in front of me.

Captain Jim Parker, my instructor pilot in the front seat, is maneuvering the aircraft erratically, cavorting around the heavens in a random series of violent, chaotic swoops, loops, and swerves that will disorient me and leave me not knowing which way is up. And he is succeeding. I have no idea whether my rear end is pointed to the sky or in some other random direction. So much for flying by the seat of my pants.

"Willard, you have the aircraft." Over the intercom from my I.P.

I have all kinds of gauges and instruments in front of me. They are supposed to tell me altitude, speed, direction, rate of climb or descent, what the aircraft's attitude is—right-side up, upside down, sideways, going up, going down, or some diabolical combination of everything—and what the engine is doing. Is it at idle? Important to know. If the I.P. has left us climbing or turning at low speed, it will be easy for me to stall us out and put the plane into a spin.

But maybe he has left us diving straight toward the ground at high speed and full throttle. Or upside down and in a tight turn. You would think the seat of my pants would tell me, but I have no idea. He has left me feeling like I am flying straight and level now. But am I? Or is it an old I.P.'s trick to drive his student into an irredeemable state of confusion?

Instructors are good at this. It is their job: Weed out the ones who get confused. Discover those who get airsick. Flunk those whose nerves aren't up to it.

"You have the aircraft" rings in my ears.

I am fixated on the confusion of instruments in front of me.

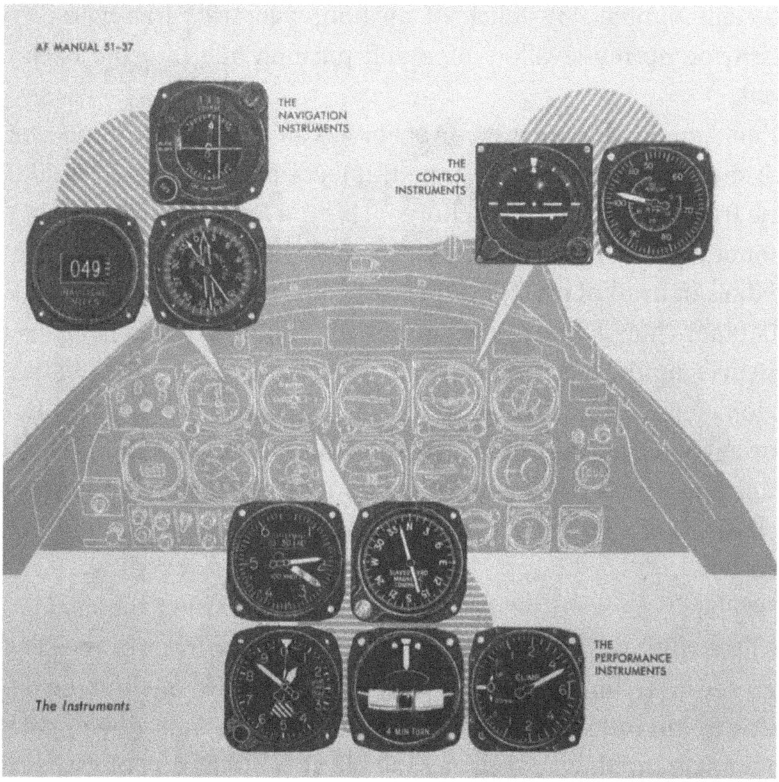

There are so many. They are different shapes and sizes with different kinds of indicators. And I have to interpret each one accurately and fast. If I am wrong or slow, and the I.P. isn't quick enough to rescue us...

I remember the preflight briefing: "Willard, if I say 'Eject!' and you say 'What?' you'll be talkin' to yerself."

I try to look at everything at once, but it doesn't work. It is visual noise. I can't interpret what I am seeing. *Calm down. You can handle this.*

The attitude indicator is a gyro-driven instrument with a miniature airplane replica at its center. It shows the aircraft's position relative to the horizon. I stare at it. The little airplane is upside down and tilted, and way below the horizon. I still feel like we are flying straight and level. But this thing says we are upside down, turning and diving. I flick my eyes to the altimeter. The needles are unwinding like the hands on a Dali clock in a drunken nightmare. The pointer on the airspeed indicator is going the other way. I glance at the engine RPM: 100%!

I gotta do something fast, or we're gonna be the world's biggest, fastest post-hole digger. They talk about "augering in." Now I know why.

I pull the power to idle, throw the stick to the left and aileron-roll the aircraft, so the little wings of the image on the dial are parallel with the artificial horizon, then pull the stick back until the tiny plane on the gauge says we are in level flight. Next, I check the velocity, add a little power to steady the airspeed, and try to keep us flying straight and level according to all the instruments.

It sure doesn't feel straight and level to the seat of my pants. I could swear I was now in a climbing turn. It is all I can do to believe the instruments and not my derriere. Hard to keep the gauges where they belong, but I do it, more or less. That is what I've been taught.

"Well, I had my doubts there for a minute, Willard, but well done. You can peel back the hood now. I'll buy you a beer when we get back, that is, if you can get us on the ground."

Parker knew how to get the best out of me without me getting us killed and knew how to make me feel good afterward. I won't forget him.

And a beer never tasted so good.

RVWG Meeting, April 10, 2021

Prompt #1: The loss of innocence. Veterans talk about going off to war, losing their innocence, and then longing to regain it. Is going off to war like losing your virginity? Are we forever soiled or transformed by it? After returning home from Vietnam, Afghanistan, or Iraq, when did you sense your loss of innocence, and, at what point after being home, did you long for the return to innocence?

Innocent Abroad

By Charles F. Willard

Is going off to war like losing your virginity? No. I recall that "loss" as pleasurable. But I lost innocence, oh, yes. And it didn't take long.

I am a country boy, born, bred, and at heart. You know. Think Gomer Pyle, Forrest Gump, Barney Fife–a real rube. But the birds and the bees were no mystery to me. I took care of our animal herds from the time I was six, and animals have no inhibitions about sex, especially billy-goats—no loss of innocence there.

My loss of innocence came quickly when I entered the military. I found too quickly how life <u>really</u> works. My country/small-town upbringing said, "Work hard, do right, use caution. Everything will work out." And that's the way it was through college. Life seemed both fair and near-eternal. Like most young people, I felt nigh indestructible and overconfident in my evolving skills and smarts, knowing I would avoid all pitfalls and peril. Because I always had. "Naivete triumphs uber alles."

Attitude adjustment came early and often. It didn't take long to learn that military service is often a dangerous and perhaps terminal profession.

Vignette #1: Early in pilot training, I nearly died from hypoxia (lack of oxygen). My helmet and oxygen mask didn't fit right, and I passed out in the middle of a maneuver. My alert instructor saved me. It would have been a pleasant way to go... just drift off.

Vignette #2: On my last training flight in a T-33, we came within feet and milliseconds of a fiery mid-air collision when another aircraft invaded our assigned route and altitude.

Vignette #3: Three of my four pilot training instructors died on the job in air crashes.

Vignette #4: Still in training, this time in C-130s, our instructor alerted us in the middle of the night to fly a secret mission across the country and back, about 18 hours of flying. Then, with hundreds of other aircraft, we sat on alert at Homestead AFB in Florida, waiting to drop troops into Cuba. It was the Cuban Missile Crisis, and they were predicting that 80% of us would be shot down. Thank God Khrushchev turned that ship around.

Vignette #5: I got word that two of my pilot training buddies, Ray Doyle and Dick MacDougald, were killed in separate air crashes in Southeast Asia. Others followed—too many names on the granite wall.

You get the idea. Loss of innocence. Remorse, guilt, "Why not me?" I don't long for the return of innocence, but it sure would be nice to have all those guys back.

Death in the Afternoon

By Charles F. Willard

In the early 1960s, Air Force pilot training at Williams Air Force Base in Arizona was divided into two six-month segments with two different aircraft types—and they were very different.

Captain Jim Ellis was one of my instructors in the first six-month part—Primary Pilot Training. Here, we flew the twin-jet Cessna T-37B "Tweety Bird." It was a placid, forgiving little sweetheart of an airplane. "Your grandmother could fly this airplane, Willard, so don't screw up."

Captain Ellis was a family man. His two little daughters came running when he walked through the door. Paying no mind to his sweat-stained, salty-backed flight suit, they wrapped their tiny arms around him as he squatted to greet them. His pretty young wife, dark tendrils of auburn hair curling around her cheeks, wiped her hands on her apron and puckered her lips for him to kiss as he rose.

The scene reminded me of my dad coming home from his traveling salesman job at the end of the week. My little sisters would run to him that way, so glad to see him, while my mom stood smiling quietly in the background, waiting while the girls had their turn.

Capt. Ellis had invited us, his three pilot training students, home for a Friday night beer and cook-out after a long week. That was a rare occurrence in pilot training. Most instructors didn't fraternize with their students, preferring the company of fellow instructors at the Officers' Club bar on a Friday night after work.

But Jim didn't frequent the bar much. Instead, off-duty usually found him home with his family.

He was a strict, no-nonsense instructor, so this invitation caught us off guard. And the tenderness he showed his family opened our eyes a little. The tough guy had another side. He was a different person off-duty, and that night we got to know him as a person. He inquired about our families and backgrounds, asked how we ended up in pilot training, and wondered if we had any problems with the course that he could help with. It seemed that he genuinely cared about us. It was a shock.

We gradually relaxed and ended up having a great time that evening.

But on Monday morning, the mask was back on. Ellis reverted to "tough guy" in the cockpit. Exacting, intolerant of any error or slip-up, never letting up on the criticism, riding our black-and-blue derrieres through every maneuver, praising only grudgingly when we did well. In the air, we could not relax for an instant.

Looking back, it was what needed to be done. Flying is an exacting art; a microsecond of inattention can kill you. A speeding aircraft will not wait around while you get your act together.

So Ellis prepared us well. When the check rides came, we knew what to expect. We did not relax; we did not let up. We passed and left Captain Ellis behind, instructing the next class of Primary students while we went on to Basic Pilot Training in the Lockheed T-33.

****

Compared to the serene Tweet, the Lockheed T-33, known as the T-Bird, was a crotchety, temperamental old biddy. She was the direct descendent of the P-80 "Shooting Star," America's first jet combat aircraft, hastily designed and built during World War II (in 143 days!) by the legendary Kelly Johnson of Lockheed "Skunk Works" fame. The hope was that she would counter the German's Messerschmidt 262 jet. Given her quick creation, there was no time to refine her character or send her to finishing school.

She was what she was, and given these circumstances, her behavior under certain conditions could be disagreeable, verging on malicious. Her controls and instruments were primitive compared to modern aircraft. She had plenty of idiosyncrasies, quirks, and peccadillos.

In learning to fly her, there were certain things you should not do. If you did them, she could give you a vicious bite. For instance, you had to get the fuel-ignition sequence correct, or there would be an explosion in her rear. If you pushed her too hard or fast in a dive, her controls would freeze, and she could plant you deep in the ground.

And you had to keep the engine RPM high on the final approach to landing because even when you shoved the throttle to the wall expecting an instant surge of power to get you out of a tight spot, it took six or seven seconds for the engine to spool up. That is six or seven forevers if you need power desperately, so you'd probably complete the flight by plowing up that cornfield half a mile short of the runway. And besides, if you rammed her throttle quickly forward, you probably over-temped her engine and blew its turbine blades through the sides of the fuselage.

My instructor in the T-Bird, Captain Jim Parker, was a seasoned T-33 pilot who made us fully aware of all of these eccentricities and, like Jim Ellis, brooked no nonsense in the cockpit. So I survived the T-Bird and went on to a successful flying career; i.e., I walked away from all my landings.

But still, teaching students to fly high-performance aircraft is risky business. Instructor pilots probably experience more "microseconds of sheer terror" than all the rest of us pilots combined.

****

It was a shock when I heard. I was flying C-130s by that time, worldwide missions into remote areas with rugged terrain and all kinds of weather conditions, challenging situations requiring high skill and intense concentration. However, the disciplines I had absorbed in pilot training were serving me well.

I learned of the accident in a letter from one of my pilot training buddies. Ellis had recently transitioned from the T-37 to instructing students in the T-33. After hundreds of hours in the "Sweet Tweet," perhaps he found the transition to the T-Bird as initially unsettling as I had—sort of like dating the Sweetheart of Sigma Chi for years before ending up with the Vituperative Vixen from Valhalla.

Later, I pieced together what had happened.

It began like thousands of other training flights. Ellis and his student briefed the mission, grabbed their chutes and helmets, trotted to the flight line, pre-flighted their aircraft, and took off into the sunny afternoon sky.

They practiced maneuvers in the training area, then returned to the traffic pattern at the base to shoot some touch-and-go landings before the lesson period ended. This is a *routine* exercise, and by the time a student completes pilot training, he has completed hundreds of these. Instructors, of course, have done them thousands of times.

Routine. Routine is a bad word in flying. Once you regard something as routine, you begin to relax. When you relax, your concentration wavers. Maybe your mind is not staying "ahead of the aircraft," and once you get "behind the aircraft," you rarely catch up. Things move too fast. Maybe that is what happened. A typical T-33 landing goes like this: We begin our approach by flying parallel to the runway on the "downwind leg." At this point, we are tooting along at about 200 knots, flying 1500 feet above the ground, in the opposite direction to the way we will land (See the biggest airplane in the lower-left corner of the diagram).

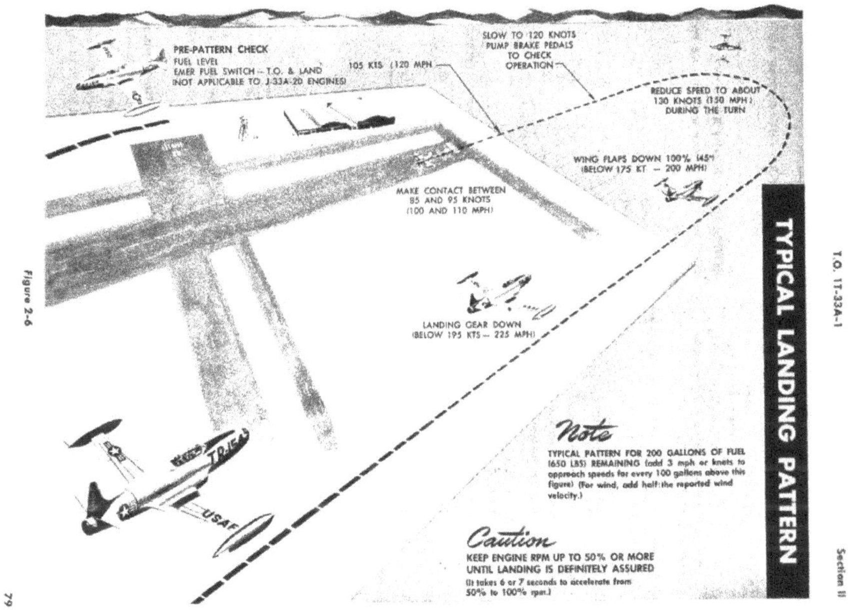

Here, we run the "before landing checklist" to get the airplane ready for landing. We lower the landing gear and check that all three wheels are down and locked (three in the green). The extended landing gear's drag (air friction) helps slow us down. Now we lower the flaps. These give us some additional lift at lower speeds, but they also increase the drag on the airplane and slow us further. We add some power to keep our airspeed high enough to make the steep, descending 180-degree turn that will line us up with the runway for landing.

This turn is the trickiest part of our landing pattern. It requires skill, concentration, situational awareness, presence of mind—and instantaneous reaction to all that is rapidly changing in front of us. We are living in the middle of a high-speed movie.

It is our point of maximum danger. We are simultaneously slowing down and making a steep, descending turn, an ultimate test of pilot skill to keep it all under control.

As we enter that fateful turn, if our initial bank is not steep enough or our airspeed is too high (or our tailwind is too brisk), we might not be turning sharp enough to line up with the runway. So we will need to tighten our turn: Steepen our bank, and reduce power simultaneously, a potentially deadly combination that puts us in grave danger of stalling the airplane. A "stall" means there has ceased to be enough air flowing over the wings to keep them flying. And we are now too low to recover from a stall before we fall out of the air and crash. Even worse, we are too low to bail out. Our parachutes will not have time to open before we hit the ground.

Here is an excerpt from the accident report.

> *The board determined from the evidence that the student was flying the aircraft in the final turn for a touch-and-go landing. While maneuvering at low altitude, the aircraft went into a descending left turn, with the bank increasing steadily beyond the 45-degree maximum allowable in a final turn. As a result of the excessive bank angle combined with a final turn airspeed that was too low, the T-33 stalled, departing controlled flight with a rapid descending roll to the left. Because of the low altitude of the stall, the instructor pilot could not successfully recover the aircraft before crashing. Neither crew member survived.*

The instant before impact

Years later, I rediscovered this verse in my pilot training yearbook:

The Instructor's Prayer

The Student is my pilot, him I shall not want.
He maketh me forced landings in cow pastures,
He leadeth me into trees and high-tension wires,
And into deep waters,
He tortures my soul.
Yea, since I ride through the air in the shadow of death,
I shall fear all evil, for he is with me.
His stick and his rudder confuse me,
He destroys my confidence.
My temper runneth over.
Surely goodness and mercy have followed me
all the days of my life
And I shall be grateful if you will spare me, Lord
and let this yo-yo graduate.

This time the prayer went unanswered.

Night Terrors at Flight Level 3-3-0

By Charles F. Willard

We are nearing the end of an exhausting mission, two weeks of nearly non-stop flying, day and night, with minimum rest. I am the still-green second pilot on the crew of our Air Force C-130 Hercules, just one of many aircraft shuttling supplies and equipment from American bases in Southeast Asia–Clark in the Philippines, Andersen on Guam, Naha and Kadina on Okinawa–into airfields in Thailand and Vietnam. It is early in 1963. The initial U.S. build-up in support of the South Vietnam government against the communist insurgency from the North is beginning, and we are doing our part.

At last, our tired crew is on its way home across the Pacific to Charleston Air Force Base on the east coast of the good old U.S.A. It is a long, 11,000-mile slog in a C-130. The airplane is built for hauling large loads of stuff into remote, primitive airstrips, not for speed. We leave Danang, with stops at Guam and Midway Island for fuel, then on to Hickam AFB in Hawaii for crew rest. No time to enjoy Waikiki Beach, and we are anxious to get home. It is nearly 5,000 miles, and we normally need a refueling stop. The official range of a C-130 is listed as 2,361 miles, but we have no cargo to weigh us down, and the forecast winds are favorable. So we decide to try for a non-stop C-130 distance record. By the route we are assigned to fly, it is 4,823 miles, 15 hours, and 41 minutes to Charleston.

Leaving Hickam with a full load of fuel, we put the wheels in the well at 1400 hours local time (2 in the afternoon), and lumber out to sea past Diamond Head, climbing slowly toward our initial altitude of 18,000 feet. We will step-climb all the way home as we burn off fuel, ending up near the C-130's service ceiling of 34,000 feet or, in aviation parlance, Flight Level 3-4-0. At that altitude, three-quarters of the earth's atmosphere is below us. Up that high, there is much less resistance to our bulky airframe, so the higher we climb, the less fuel we will burn.

By midnight Pacific Coast time, the outline of the bright-lit U.S. mainland begins to take form. It is a crystal-clear night, and from our vista at thirty-one thousand feet, nearly six miles in the air, the vast, twinkling display of the San Fernando Valley and the Los Angeles Basin sparkles below, a welcome sight after so long over the ocean. Above land now, I request a further climb to our final altitude, Flight Level 3-3-0, from Los Angeles Center.

Request granted, I give up my position in the left seat to our aircraft commander, Captain Dick Davis. I have been sitting here for nearly eight butt-numbing hours, and it's my turn for a nap. I climb into the upper crew bunk on the back bulkhead of the flight deck, stick wax plugs into my ears against the fatiguing 98-decibel roar of our four turboprops, and fall quickly asleep, dreaming of home....

Something incongruous pulls me out of my sugar-plumb slumber. It penetrates the enervating drone of the engines and brings me groggily to the surface. Some sort of a thud, I think. Then, in a moment, what sounds like a muffled explosion. Immediately the aircraft begins to shake, a violent, trembling shimmy. This is not supposed to be!

Fully awake now, I look down and see Captain Davis with both his hands gripping the control yoke, which is oscillating rapidly forward and back despite his efforts to steady it. I can feel the aircraft pitching and bucking like we are in severe turbulence. A glance out the cockpit windows shows a clear, calm sky, bright with moon and stars.

Panic.

I scramble from the bunk and grab the nearest headset to discover what's happening.

I recognize the tense but steady voice of Captain Davis. "Fort Worth Center, MATS 21786 at Flight Level 3-3-0."

"MATS 21786, Fort Worth Center, good morning."

"Fort Worth, 21786. We have a control problem. I'm declaring an emergency at this time. We are unable to maintain Flight Level 3-3-0. I've applied full power, but we are losing altitude. Request clearance down to Flight Level 2-9-0."

"21786, you are cleared for descent to Flight Level 2-9-0. State the nature of your emergency."

"Fort Worth, 786. We heard an explosion followed by extreme shaking of the aircraft and a wild oscillation of the control yoke. We may have some sort of mechanical failure. Request vector to the nearest airfield with emergency equipment. Can you confirm that there is no clear air turbulence reported?"

"786, Fort Worth. No CAT reported. Tinker Air Force Base is the nearest field, 80 miles to your northeast. Would you like a vector?"

"Roger!"

"786, turn left to heading 0-6-0, descend to Flight Level 2-1-0."

"Heading 0-6-0, descending to 2-1-0. Roger."

Then on the intercom, "Willard, are you up?"

"Yes, sir!"

"Dig out the approach plate for Tinker."

"Yes, sir." I scramble to the map case and sort through it with shaky fingers. "Here it is."

"Keep it handy. When the time comes, read it off to me when I need it. Gordon, you take over the radios."

Gordon was Gordon Almquist, the 1st pilot. He is in the right seat.

Big airplanes have a big crew for a reason. Everyone has his special job to do during each part of the mission. We have all spent countless hours training for emergencies, but we have never trained for one quite like this. As Aircraft Commander, Davis was doing his part. His first job is "Maintain Control of the Aircraft." And the second is "Crew Coordination." He is doing a superb job of each—flying the airplane and directing each crew member through this strange and scary emergency.

"Loadmaster, what can you see in the back? Anything out of the ordinary? Can you see anything when you look out the portholes in the paratroop doors? Willard, go back in the cargo compartment and help Taylor with the look-around. Check the ramp and door and see if you see anything coming apart in the tail section."

I go back. The craft is shaking so violently that it is hard to maintain my footing. Taylor, the loadmaster, and I do a thorough inspection of everything we can see. Nothing visible, inside or out. But we cannot see the tail surfaces through the small portholes. I scramble back to the flight deck and report.

"No, sir. Nothing visible at all."

"Edkins," Davis says, "Stick the sextant through the port and see if you can see anything going on with the tail."

Dave Edkins is our navigator. The sextant is an optical navigational instrument with a small periscope. The navigator can see the sky when he pokes the periscope part through a small port in the plane's roof. He can tell the aircraft's location by taking positional readings from several known stars, just like the sailors of old. This is the early 1960s, before navigational satellites and GPS. Using the sextant is one of the only ways to tell where we are when we are over the ocean.

Now, Edkins is using it to see if there is any damage to our tail surfaces.

"It's pretty dark out there, sir. From what little I can see, everything looks normal. But I can only see the vertical fin. The horizontal tail surfaces are too low for me to see through the sextant."

"Dammit! Well, something bad is going on back there. It feels like some sort of structural failure. I hope we're not losing the hinge pin in the elevator. And it's hard to believe the main tailplane spar would fracture. If any of those things are happening, we might be in real trouble."

It was like someone had stuck an icicle down the back of my flight suit. Four miles in the air in the middle of the night, and the airplane might be coming apart. And we carry no parachutes. There is no way out except a successful landing–one we can walk away from.

"I'm going to take us through some controllability maneuvers. Everyone, strap in tight. We have to see if we can maintain control as we slow her down to landing speed."

Davis gradually pulls back on the throttles, and the aircraft begins to slow. When we reach 150 knots, he asks Gordon to slowly lower the flaps. The speed diminishes, and the vibration is less violent, but the airplane begins a wallowing corkscrew motion. My stomach churns in time with the sickening ballet.

It is taking all of Davis's skill and coordination to counteract the unrelenting gyrations of the yoke as he struggles to keep the plane on some kind of an even keel.

"Well, I don't like that. She's making me work. Now the big test. I'm going to lower the gear and slow toward landing speed. Hang on tight."

You can bet we are hanging on tight, with white knuckles and clenched teeth.

"If she falls off on a wing, I'll have to dive her to get her flying again. I hope I can keep her from falling into a spin."

Oh, no, I think, *a 70-ton corkscrew!*

At Davis's direction, Gordon pulls the gear handle down. As the wheels begin to lower, we can feel the buffeting increase as she slows, wobbling toward touch-down velocity. She is barely flying now, almost on the edge of a stall, mushing and yawing. Davis is using every bit of his "seat-of-the-pants" skill to keep her airborne.

"All right. I think I can land her when we get to Tinker. Gear up, flaps up," says Davis, as he pushes the throttles gently forward, increasing speed again.

Without realizing it, we've all been holding our breaths. A group exhale is audible over the intercom. Then we realize what Davis said: I *think* I can land her.

So, the fear that grips us subsides only a little. She might handle quite differently in the denser air at the surface. If she doesn't come apart in the air first. Old pilot's adage: *It's better to be down there wishing you were up here than to be up here wishing you were down there.*

We still have to get her safely on the ground. Our collective fate rests not just in Davis's hands. It will require our best "crew coordination." Let's pray that, whatever happens on the way down, we all know what to do.

"There is one more thing I want to try," Davis says, "It's possible we had a life raft ejection."

The C-130 carries four large rubber 20-man life rafts. These are folded, compressed, and stored in compartments just below the rear top surfaces of each wing. Each raft has a bleed air valve designed to leak excess pressure as the aircraft climbs into thinner air. This keeps them from expanding and blowing the doors off the compartments as the airplane reaches higher altitudes. If one of these valves has failed....

"If a life raft blew out, it could have been caught in the slip-stream and somehow lodged back there on the tail. I'm going to yaw the aircraft, and if that is the problem, maybe the yawing will dislodge it."

Gently at first, Davis pushes on one rudder pedal and then the other. Nothing. Then he gets more forceful. The plane slues and corkscrews crazily, on the edge of control. Bile rises in my throat. Finally, the craft steadies as Captain Davis eases off the rudder pedals. But the shuddering continues with no noticeable change.

"Nuts. So much for that brainstorm. I was hoping that was the problem. But, if it is, she sure is stuck there. We'll have to risk bringing her in at Tinker. Gordon, you handle the radios. Willard, you read off the checklists. I gotta concentrate on getting this baby on the ground."

She continues to oscillate as we ease our way down. Mushier and harder to handle as we get lower and slower. Even in a normal airplane, the landing approach is always the most critical point of any flight. Airspeed is bleeding off. Engine power is low. Altitude is deserting you fast. The margin for error is razor-thin. *The most useless things in aviation: the altitude above you, the airspeed you just lost, and the runway behind you.*

Tinker Control gives us a radar vector direct to final approach. The weather is clear and calm, and a beautiful spring dawn is beginning to break, purple, red, and yellow in the east. God, I hope it won't be our last.

The approach lights strobe up at us, leading us to the strip. We can see the flashing beacons of the crash trucks lining the tarmac. The shaking, yawing, and mushing are reaching a violent crescendo. Will it pitch us into the ground before we get her low enough to land? Strapped in tight against whatever happens, sweat is pouring down my forehead and into my eyes, and I am shaking all over, not sure if it is only because of the aircraft vibrations.

Carrying more airspeed than usual as a safety buffer, Davis prangs us onto the ground, hot and firm. She bounces once and settles into a screeching rush, screaming past the crash vehicles as they scramble from their positions and begin to roll after us.

Davis and Gordon both have their feet pressed full on the brakes. We can feel the chatter of the anti-skid as the fat main gear tires grab the concrete and begin to smoke. Davis hauls the prop levers into full reverse, and we can hear the banshee howl of the engines as they do their work.

Relief washes over us as we barrel to a stop in the middle of the long runway. The squadron of firetrucks, ambulances, and assorted emergency vehicles that raced after us as we careened down the runway now surround the airplane. The profusion of flashing red lights gives a comforting welcome to the safety of the ground. But, lucky for us, their services aren't needed.

"MATS 21786, Tinker Tower. Good to have you safely on terra firma. The Fire Captain reports that a large rubber life raft fell off your right horizontal stabilizer as you rolled out."

"Roger, Tinker. Thank you."

Well, the distance record and getting home will have to wait a little longer, but we've cheated death again. Later, we get an Air Force "Good Show" for saving ourselves and the airplane.

Our crew inspects the repaired life raft compartment, April 1963

Dropping Troops

By Charles F. Willard

One of the most enjoyable sensations in aviation comes during low-level flying. Careening over the terrain just above the treetops, lower than the surrounding hills and protrusions, dodging among the obstacles, delivers a heightened sense of motion. It launches an immediate, sustained adrenaline rush.

We practiced dropping paratroopers a lot during my time flying Air Force C-130s in the first half of the 1960s. It was fun for us, not so much for the troops. It usually went like this:

Early on a South Carolina summer morning, we take off from Charleston Air Force Base and practice our low-level flying all the way up to Pope AFB in North Carolina. We skim low over the forested swamps of the Carolina lowlands, watching the wisps of smoke come up through the trees into the sultry-warm morning air. This is not countryfolk heating their cabins in the woods. The temperature is already in the 80's. The smoke is from Pappy, Bubba, and Billie-Bob brewing their white-lightning moonshine. The Revenuers are ground-pounders, but what they really need is a chopper to spot the goings-on.

Thermals begin to rise off the sun-heated land, and the turbulence starts. We are not far above the ground, so it is jarring, like a bobsled cayuting downhill on a rough-ice run. The shock-mounted instrument panel blurs in front of me. I see three of every gauge, and the pounding wrenches us around against the belts and harnesses that secure us to our seats. The roar from the turboprops and the vibrations from all the fittings pound our eardrums, amplifying the experience.

We approach Lake Moultrie, 250 feet over the water, 250 knots airspeed. We can see fishermen in their boats, casting their lines, hoping to snag breakfast. At this speed and altitude, they won't see us or hear us until we burst over their heads like a thunderclap. For some of them, it may be a jeans-wetting, pants-pooping moment. The thought gives us a chuckle. It's a mean, heartless, giggle-inspiring trick.

We land at Pope, adjacent to Fort Bragg Army post, and onload the paratroops we will be dropping, a full complement of 64, in full paratroop regalia. Parachutes, of course, main and reserve, plus personal weapons and additional soldier paraphernalia secured to them by belts and cords. They are overloaded and uncomfortable as they strap into the aluminum-framed, canvas-bottomed, web-backed, orthopedically offensive seats that line the sides of our cargo bay. For some, this will be their first jump. They are shaky, nerves on display, sweating through their thick, protective clothing, perspiration running from under their heavy helmets as they settle into the confined, overheated space.

Burned into my memory is a night back in October of 1962. I was a green student in C-130 training, and my roommate and I were alerted after midnight by Major Bob Byers, our flight instructor. There was a national emergency. The country was suddenly on a war footing. It was the Cuban missile crisis. We sat in our C-130 on the ground at Homestead AFB in Florida, ninety miles from Cuban shores, waiting to be told we would drop paratroopers over Havana. They said to expect 80% casualties. Then Nikita turned that ship around, and we were saved.

Today, hopefully, there will be no losses, although shattered ankles or fractured femurs or fibulae aren't unheard-of, and once in a great while, a chute fails to open. Not today, please.

We crank up and get going. For those in the back, it will be a hell of a ride. We need to stay under the enemy's radar and get in and out fast. It will be a loud, uncomfortable, jarring/jouncing ride for the boys in the back. If the combination of that and their nerves don't give them the upchucks, we'll add the piece de resistance at the end: a pop-up maneuver.

Here's how it works: We are tooting along, fast and furious, through the considerable turbulence at 250 knots, 250 feet over the ground, approaching the drop zone. We can't drop troops from 250 feet. Not enough time for their chutes to open. And not at that speed. The opening shock might tear the silk from the shroud lines. So today, we will drop from 1,500 feet at 125 knots. That's where the pop-up maneuver comes in. It's how we get from fast and low to high and slow.

As we approach the drop zone, we pull the power to idle and yank back on the control yoke. This puts us into a two or three-G climb and gets us from 250 feet up to 1,500 feet in a hurry. It also bleeds off our airspeed to the required 125 knots. As soon as we are near 1,500 feet, we jam the control yoke forward to hold the altitude. That gives us about two negative G's. This is when the bottom of the stomach meets the roof of the mouth. Everything in said stomach most likely will make an explosive exit of the oral cavity, even through clenched teeth. Those paratroops who think fast will whip their helmets off their heads, providing a handy repository for the oral excretion. Then there will just be time to re-strap the helmet to the head before the green jump light comes on, and it is time to egress the aircraft. By this time, no one is afraid to jump. They are just eager to exit this infernal flying machine.

For us up front, just another fun day of aviating.

Russian Roulette at Rhein-Main

By Charles F. Willard

It has been a long flight from Charleston Air Force Base in South Carolina to Norfolk Naval Air Station in Virginia, then across the Atlantic to Mildenhall in England. And we still have further to go.

At the front of our minds, as we negotiate our instrument approach through soggy, gray weather into Mildenhall Air Base in England, is the near-tragedy we had witnessed while on the ground at Norfolk.

A Navy P2V sub-hunter aircraft, coming in to land after an all-night patrol, had crashed right in front of us. The pilot lost sight of the runway in a rain squall, and the aircraft slammed into the concrete from 50 feet in the air, driving the right landing gear strut through the wing and severing it from the rest of the plane. Gas from the ruptured wing tank exploded into a hundred-foot fireball as the plane cartwheeled down the runway, coming to rest a few hundred feet away. We were sure everyone on board was being burned alive.

But several huge crash wagons appeared from nowhere and pushed the flames away with torrents of foam. We watched amazed as, one by one, all ten airmen staggered from the wreckage. Fate had fired a blank this time, but the horror of the scene shook us badly, and we still had to go on and fly our mission.

After the brief stop at Mildenhall, we would fly on to "Frankfurt am Main" in West Germany. This was the Cold War, and Germany was still divided. The "Iron Curtain" ran straight through the middle from north to south. And Berlin was still separated by the Berlin Wall between the free West and the communist East. East Germany surrounded the beleaguered city. The Russians (USSR) kept up a continual agitation, always threatening to cut off highway access to Berlin (which they had done for eight months in 1948-49, resulting in the Berlin Airlift). Rhein-Main Air Base at Frankfurt was the principal American terminal for departure and return of those perilous flights into and out of Tempelhof Airdrome in West Berlin. Those flights saved the West Berliners from starving and freezing during that harsh European winter.

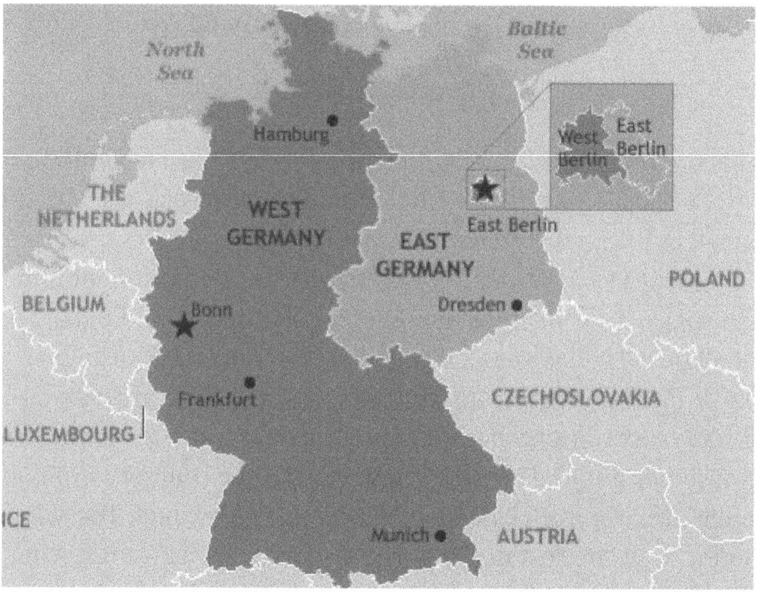

And the base remained vital to staving off Soviet ambitions. Many of our Air Force MATS (Military Air Transport Service) flights into Europe during that era were to counter the aggressive actions of the USSR. Our flight today was just a tiny cog in what was a continual moving and staging of troops and equipment around the West European perimeter of the Iron Curtain. Today, we were moving a full complement of paratroops from England to Rhein-Main—one of those "just in case..." operations.

By the time we refueled and our passengers, with their equipment, were finally aboard, it was late afternoon of a dismal, drizzly New Year's Eve, 1963. I was copilot for Sam Lewis, who had passed his check ride for Aircraft Commander only a week ago. Sam was a good pilot, a skilled aviator, and a good crew coordinator with a common-sense head on his shoulders. But this was his first flight as A/C of a C-130 Hercules aircraft and on an overseas mission, to boot. The weather was not in his favor. Most of Europe was semi-socked in with winter rain and fog, and it looked like it would be a "minimums approach" at Rhein-Main, about one-and-a-half C-130 flying hours away.

****

"Rhein-Main Approach, MATS 37794, at 2,000 feet, on a heading of 0-9er-0 with 72 souls on board. Requesting Ground Controlled Approach to Runway 0-9-er Right."

We had flown on instruments all the way from Mildenhall, and the destination weather was heavy fog with ceiling and visibility near minimums—pucker time.

A GCA controller could talk us down through the murk to 100 feet above the ground and within a quarter of a mile of the runway. If we could not see the approach lights at that altitude and distance, we would have to execute a "Missed Approach" procedure. That was something we had to plan for ahead of time. No scrambling around in the cockpit at the last minute, trying to figure it out when we were that low and slow.

Slow is a relative term. We would be moving at nearly 150 miles per hour, over 200 feet per second. Many immovable obstacles—buildings, antennas, hills—are sticking up around the runway, higher than 100 feet—nothing we wanted to hit going 200 feet per second.

"Roger, MATS 794, descend to and maintain 1,000 feet. Contact Rhein Main GCA control on 123.6." That was the VHF radio frequency on which we could talk to the GCA guy.

"MATS 37794, Rhein-Main Radar," he answered when I had dialed him up and made the call, "I have you on glide path, right of centerline. Turn left to a heading of 085 to intercept centerline." The voice of the Rhein-Main Ground Controlled Approach (GCA) technician came over our headsets, crisp and clear.

"You are now intercepting centerline. Turn right to 091 degrees... On centerline, on glide path, passing 300 feet. Report runway in sight."

We were still "in the soup." Sam was fixated on the instruments, our only means of knowing where we were, how high, how fast, how much bank, and what our rate of descent was. My job was to run the radios and the checklists and keep my eyes peeled out the windows to look through the fog for the runway lights.

"Passing 200 feet, drifting right of centerline, turn left to 085, report runway in sight."

Things were beginning to get dicey. The crosswind of a few knots was pushing us to the right of the runway. Sam was holding our airspeed at 130 knots until I told him the runway was in sight.

The unemotional voice of the GCA controller: "Drifting further right, passing 150 feet, approaching minimums, take over visually or execute Missed Approach."

Decision time.

"I have the runway!" The relief in my voice echoed in my own headset. That was Sam's signal to look up from the instruments, acquire the runway visually, and make whatever corrections he needed to land us all safely.

"I don't have it!"

Panic. Where did it go? It was there when I hollered.

What flashes through the mind in an instant like that? *Have I just killed us all? Norfolk all over again!* I saw myself at the Pearly Gates with 72 dead souls on the debit side of my ledger.

Then, from Sam, "I got it! Ten o'clock." He jammed the throttles forward and put us into a 30-degree left bank, risky when we are that low and slow, and our altitude above the ground is only a little higher than the length of our left wing. Sam juked the aircraft into an "S" turn and lined us up with the runway.

Flaps, 100."

"Roger, flaps 100%," I answered.

Ba-bump, as we touched down. Not a bad landing. Fate had fired another blank. Cheated death again.

RVWG Meeting, May 2, 2020

Prompt #1: Choose a word and define it from the perspective of your time in the service.

Feeling "Vulnerable"

By Charles F. Willard

We were at the bar in the Incirlik AFB Officers' Club, Adana, Turkey. Around midnight, the phone behind the bar jangled. The barkeep answered it, listened a moment, and, holding the receiver aloft, turned around and bellowed, "Is there a Lt. Brandt here?"

Larry Brandt was our designated aircraft commander for this mission. Without looking up from his drink, he raised a hand and hollered, "Yo."

We had landed at Incirlik that afternoon and—after the long flight from Charleston AFB, South Carolina, over the Atlantic to Lajes Air Base in the Azores, then across the Mediterranean into Turkey—were granted a 24-hour crew rest.

After short naps, we—the officers on the crew, the three pilots and two navigators—had roused ourselves enough to enjoy an excellent late dinner at the "O" club and then meandered over to the bar for a nightcap—or two.... We were looking forward to a good long sleep before our scheduled 1400 hours (2 PM to you civilians) take-off the next day.

The bartender, a swarthy Turk, waddled over to Larry and handed him the phone.

"Lt. Brandt... Yes, sir. No sir. No problem, sir. Absolutely, sir. No, sir. Yes, sir! Right away, sir."

Hearing only Larry's side of the conversation, the first thought the rest of us had was, "What the hell have we done now?"

Larry turned to us with a wry look and pronounced, "That was the Base Operations Officer. We've been alerted."

"What? Are they crazy? We just landed. I thought we were supposed to have 24 hours." It was a jumble of protests from the rest of us.

"Don't matter. They got an emergency at Peshawar."

Peshawar is a city in northern Pakistan, not too far from the southern border of the Soviet Union. At that time, the USA had listening posts, as close as it could get them, all around the USSR. There was one at Peshawar, and it had "gone dark." They badly needed repair parts. Our crew was in the right place at the right (or was it wrong?) time, and we were tapped to get the parts to them ASAP.

"Larry, are you sure we're in condition to fly?" We've hardly had any rest, and we've been drinking."

"Don't matter. They need the parts."

So, we flew.

There is an old Air Force dictum: No smoking within 12 hours of take-off time and no drinking within 50 feet of the aircraft.

Or was it the other way around? We could never remember.

RVWG Meeting, 2-3-22

Prompt #3: Is a culture of "toxic masculinity" necessary within the Armed Forces, a "necessary evil" in the name of national security, and would you characterize it as such?

Toxic Culture

By Charles F. Willard

Perhaps it was Ulysses S. Grant who first declared, "The purpose of the military is to kill people and break things."

Some people would disagree with that. They would say that the primary purpose of the military is to act as a deterrent—to defend and protect. To back up diplomacy with some teeth—to give peace a chance.

That is true, but what happens when diplomacy—an imperfect art, at best—fails, and the nation's fate hangs in the balance? Then the duty of the military becomes to kill people and break things.

If a military force is to succeed in this "last resort" mission, it must be trained to be smarter, more powerful, more skillful, and more lethal than the enemy.

Can this be done without instilling a "toxic culture"? And when does a military culture become toxic?

To succeed and survive, a soldier must learn to be an effective killing machine. For most, this is a life-altering transformation. But, is it even possible to focus this mindset solely on the enemy without having its poison spill over like a toxic waste, evermore tainting anything and anyone that gets in its way?

Does soldiers' training forever change how they regard their fellow humans? Do they now place a degraded value on humanity, including their own? When one loses respect for oneself, one loses respect for others, also.

Throughout history, armies have been groups of men raping and pillaging. Male aggression dates back to the caveman with his club, selecting and subduing his mate and dragging her by the hair back to his lair to do his bidding and bear and nurture his children.

Testosterone at work—the female, living in virtual slavery but protected by the dominance of her mate from perhaps a worse fate. The female has had to fight for equality of personhood right up to the present day, but with wiles rather than weapons.

And one ponders.... As women in the military are also taught to be aggressive killing machines, with the expectation that they will be equally lethal combatants alongside their male counterparts, will a similar toxin saturate their psyches too?

Many questions, no good answers.

Our Failed Attempt to Penetrate Soviet Airspace

By Charles F. Willard

We called it "The Peshawar Shuttle." Our mission was to shuttle supplies from the U. S. depot in Karachi up to Peshawar, Pakistan, near the Khyber Pass. It was the site of our listening post close to the southern borders of the Soviet Union. Rumor was that in addition to "listening," Peshawar was also the site of more furtive activities. Whenever we landed there, we would notice a few unmarked C-47 (the old Douglas DC-3) aircraft sitting on the opposite side of the field. Was the CIA's air arm, Air America, using these for clandestine nighttime low-level incursions across the border into Afghanistan or one of the Soviet "-Stans?" One never knew in those days.

We departed Incirlik Air Base outside of Adana, Turkey, on a hot, sunny morning in July 1965 on our way to Karachi to begin shuttling. On the way, we stopped in Tehran, Iran, to deliver a couple of pallets of equipment for the Shah's Iranian Air Force. We (the U.S.A) were still good friends with Iran then. The Islamic fundamentalists didn't overthrow the Shah until 1979.

We offloaded, refueled, and were soon on our way east into the bright afternoon sky. I was in the left seat for this leg, about 1,200 miles to Karachi. Everything was humming along nicely. Our navigator gave me the heading for the next leg, and I set it into the autopilot. The flight engineer was running the overhead panel—fuel management, electricals, HVAC, all that stuff. I was monitoring the engine and flight instruments, all in the green, so I was essentially unemployed up there at 25,000 feet.

I had my feet propped on the footrests that protruded from just below the instrument panel, there to relieve the pilot from an hour-after-hour, feet-on-the-rudder pedals posture. I had the airways map on my lap and was looking out the window at the monotonous terrain, attempting to locate our position from the scant terrain features shown on the chart.

There was a large lake down there, in the middle of the desert. I looked down at my chart. Yep, here was the lake. By its shape, that must be the one I'm looking at. Wait a minute! On the chart, that lake should be on our left. But it's not! It is way off to our right.

"Navigator, come up here a minute, and look out the window." I pointed out the discrepancy.

"Oh, my God! Turn right 90 degrees! We're about to fly into Turkmenistan!"

I cranked the autopilot over into a steep right bank. Soon we were back on course. Navigating is not an ideal profession for someone who has dyslexia. Turns out our navigator had given us a course of 103 degrees when he had meant 130 degrees. No harm done, and we continued on our way. But a few more minutes and the story could have ended differently.

Addendum: During the Cold War, at least 30 U.S. aircraft were lost on spy missions over the Soviet Union. We know about the U-2 and Francis Gary Powers, but most of these losses were not revealed until after the fall of the USSR in the early '90s. At least one was a C-130, "Air Force 60528." It is memorialized by a replica near the NSA Headquarters gate at Fort Meade, Maryland. Seventeen crewmembers died when a MIG-17 shot down their aircraft over Soviet Armenia.

But most of these losses were secret. They were never revealed nor acknowledged, nor were any crew remains returned. We called it "The Cold War," but it was both hot and final for some of our cold warriors.

The Trek to Thule

By Charles F. Willard

I. "Grease Job" at Goose

I don't know why, but many of our C-130 missions seemed to begin in the middle of the night. They often ended the same way, usually with a hairy 3:00 AM landing in lousy weather at some God-forsaken outpost in a remote corner of the planet.

That was how it went on our hasty and unforeseen mid-winter trek to Thule Air Base, America's Ballistic Missile Early Warning Site (BMEWS), 700 miles above the Arctic Circle in the far north of Greenland.

The purpose of the Thule (pronounced "Tooley") site, along with a similar one in Alaska, was to give the first alert in the event of a nuclear launch from the Soviet Union. This often seemed a dire possibility at the height of the Cold War arms race.

The giant antennas of Thule's state-of-the-art radars could literally see beyond the curve of the earth and look deep into Soviet airspace. This visibility gave the United States a 15 to 20-minute window to launch a counter-attack. The Ruskies knew we were watching. The theory was that our ability to hurl an immediate counterstrike, assuring "mutual destruction," would lower the risk of such a "pre-emptive first strike" from the USSR.

Winter travel to Thule was generally discouraged. It sits on a wind-swept, often snowbound sub-zero reach of tundra. The winter weather is vicious and unpredictable. The situation can change from clear to a zero-zero blizzard within a few minutes, and that "phased-in" condition might last a few hours or a week. So an inbound aircrew better have enough fuel to reverse course and beat it back from whence they came. The alternative—well, there wasn't one.

This mission began early on a freezing-cold late-January morning at Charleston Air Force Base in South Carolina. A quarter-inch of snow, rare for the area, had fallen late the previous evening. As we began our preflight, the temperature was still a degree or two below freezing, and the sun, which would have made short work of the measly coating, was still about three hours below the eastern horizon.

I pop the aircraft's upper hatch, stick my head out and shine my flashlight over the top of the fuselage and out along the wings. A hoary coating glitters back at me.

I shout down to our flight engineer. "Call the deicing truck." That hoary frost on the wings would disrupt the airflow we need for lift-off, resulting in a short flight with a fiery conclusion.

While the truck sprays us down, we complete the rest of our preflight, and then we are off into the darkness; first stop at McGuire AFB, New Jersey, to pick up our cargo. Here we discover the serious nature of our mission.

A critical piece of electronics in the Thule BMEWS system has failed, and a sector of our early warning system is temporarily blind. The replacement parts they keep on hand for such emergencies have already been used. If the Soviets launch an attack from that blind sector, the USA will not know in time to respond. So, this is a "rush-rush, hush-hush" operation. Those parts need to get to Thule fast.

We get the cargo pallets loaded, along with plenty of cold weather survival gear, and depart northward into a dull, overcast early dawn, climbing our way through the thick clouds, bound for Goose Bay, Labrador, about 1,200 miles (four hours in a C-130) to the north, where we will make our first refueling stop.

****

With a solid undercast below us, we churn our way over New England, up over New Brunswick, across the Gulf of the St. Lawrence River, and into Labrador. As we start our descent, Goose Bay Approach Control gives us the weather. It will be a VFR (visual flight rules) daylight landing, but there will be nothing routine about it. They are experiencing a vicious crosswind.

As I maneuver down final approach, we are tossed around like a paper kite in the turbulent, blustering gale. It is gusting 30 to 40 knots, far above the 25 knots maximum recommended for a C-130. With its wing positioned high on the fuselage, the wind can get under it and flip the airplane. Usually, we would not land in these conditions. But, on this mission, we don't have a choice. I can see the runway now, and it is covered with patches of ice and packed with snow.

I am flying almost sideways now, struggling to keep the path of the aircraft on the runway heading as she bucks and pitches like a bronc at the Tucson rodeo. The nose is skewed 30 or 40 degrees into the crossing gale. I decide I must try landing this way—almost sideways. Of course, if the runway were bare, it wouldn't work. Even so, it is a gamble—I am gambling the airplane and possibly our lives because our orders are to get these parts to Thule.

Carrying extra speed down final approach because of the gusting wind, we come screaming in. It's a "grease job" landing, touching down soft as silk on the slick surface. We are still sideways, hurtling askew down the icy runway. Now I have to get this 70-ton bobsled straightened and stopped before we slide off the edge or hit the overrun. If I can't, it will tear the airplane apart. There is no margin for error. It will be so easy to careen into a fireball of oblivion.

Using full rudder and yoke hard over into the wind, I reverse the props on the forward wing and apply a little power to the two engines on the trailing one. I tentatively ease the nose around straight, then full-reverse all four props and jam on our anti-skid brakes. The propellers have become "repellers," the anti-skids chatter away, and we slither to a stop before the end of the runway.

We've "cheated death again," but there are probably several sets of soiled shorts among the crew.

I often marveled at the courage of our crew members, especially those who were non-pilots. They had a blind and trusting faith in us, that we knew our jobs and had the skill and presence of mind to deal with any situation. As a pilot, I was nervous if I were not the one flying the aircraft. I knew all too well the things that could go wrong—bad weather, mechanical malfunction, or pilot error—and in some situations we flew into, enemy fire. Here's to those who dare–especially when things are outside their control.

We taxi in behind the "Follow Me" truck, all four of his wheels girded in tire chains. He leads, crawling slowly toward our refueling spot, guiding us into a 90-degree turn in front of a huge hangar. The frozen ramp has a slightly downhill pitch. I can feel my tires begin to lose their grip as we start the turn, the gusting wind helping to push us slowly down the icy slope. I watch helplessly as our wingtip comes closer and closer to the looming hangar. If the wing smashes into the hangar, it will be not only an embarrassing accident but also an untimely end to a critical mission (and probably this pilot's career). Luck and the Lord are with us again, and the sliding ceases with the wingtip not ten feet from disaster. Way too close for comfort.

II. Sneaking into Sondrestrom

We use the refueling time to grab a bite at the canteen and detail the flight plan for our final two legs of the mission. We are now classified "Secret." Goose Bay departure control is not to know where we are going, and our destination will not know we are coming until we identify ourselves with an ersatz callsign on final approach. When we finish the flight plan, the wind has eased a bit, permitting a manageable takeoff.

Our next stop will be a remote outpost less than a thousand miles north, Sondre Stromfjord, Greenland, about half the remaining distance to Thule. The U.S. had initially leased Sondrestrom from Denmark during World War II. Known then as "Bluie West 8," it served as a refueling stop for our warplanes on their way to Europe. Unfortunately, some never made it and are buried in the surrounding glacial ice. Now we are using "Bluie" to gas up so we'll have enough JP-4 to retreat from Thule if the weather there is "zero-zero, and the birds are walking."

Bluie West 8 (Sondrestrom Air Base) in good weather

Sondrestrom sits near the base of a mountain, at the end of a long, steep-sided fjord. Tonight, the base is "in the soup," and an instrument approach will be required. This means we intercept a specific radio beam some twenty miles out—the same beam those warplanes used 20 years ago—and fly that beam precisely, so we don't collide with one of those craggy fjord sides. This is a dicey proposition, and when we get to the field, if it is "below minimums," we'll need to execute a "missed approach." Of course, we have to know what this is well before we have to do it. No scrambling around in the cockpit to find the correct procedure after realizing we can't land. That kind of delay, and we might end up eating rocks.

B-17 "eating rocks" at Bluie West 8

At Sondrestrom, the "missed approach" is a tight, climbing spiral to an altitude high enough to clear the surrounding terrain. Of course, you are on instruments while you are doing this, and you better hope all systems are "go"—that all your nav aids and radar (and engines) are working, so you can tell if the wind or your imprecision is drifting you blindly into one of those immovable objects.

This time, we break below the overcast about a mile out, and, runway lights in sight at eleven o'clock low, we juke the aircraft left to get us properly aligned with the strip and make a normal night landing. We do a quick top-off of the tanks and take off into the dark for the final leg.

III. The Stirring, Silent Show

Dark is what it will be from here on. We are north of the Arctic Circle now. In summer, it is "the land of the midnight sun." But it is mid-winter—"the land of eternal darkness." And there is nothing below us but snow-covered tundra and icy ocean, and nothing above us but infinite sky. It's no place for an aircraft malfunction. The surface temperature below is nearly as frigid as it is up here in the stratosphere—about minus 50 degrees.

And there are no airline routes up here, no radio beams, no navigational aids, none of those man-made systems to guide us. We will be on our own. It will be "dead reckoning." To find Thule, about a thousand miles to the north, we must "fly grid" because up here near the Pole, magnetic compasses are downright unreliable. As we approach Santa's workshop, the earth's magnetic field wanders in all directions. Our mag compass, totally confused, wanders with it. We would end up flying in circles and curly cues. So, we set our gyros, get out our grid maps, ignore the mag compass, and fly the gyro heading.

We break above the cloud cover into a vast black rotunda perforated by a million pinpricks of brilliant, twinkling light. We hang suspended in our own private planetarium. Free of the clouds, our navigator breathes a sigh of relief, pokes his sextant through the overhead port, finds Ursa Minor, and shoots the North Star. This gives him a welcome crosscheck on our location. We fly on through the darkly dazzling, freezing cold, splendidly starry night.

And then, the show begins. Dim and distant at first, a ghostly, eerie green glow rises from the horizon in front of us.

Soon we are surrounded and enveloped by—actually flying through—ethereal, undulating curtains of emerald light. They are so near and vibrant; we can almost count the photons. Softly swelling and rippling, they caress us like waves washing an ocean shore. They flow over us, leisurely and sensual. I think of elegant hula dancers, hands in graceful motion, grass skirts slowly swaying in an erotic, unhurried rhythm. Then, in burgeoning vertical waves, they tower into the ether above and stretch into the abyss of the bottomless blackness below.

We wake the sleeping members of the crew and call them into the greenhouse of the cockpit so they will not miss the stunning spectacle. And then, from somewhere in the cosmos, a silent retreat is sounded, and the scene begins to fade. The splendid, radiant fairy glow diminishes stealthily toward nothingness as though a great, unseen hand has turned an invisible rheostat steadily toward zero.

Now we are alone under the great dome of stars. Our aircraft, with its four roaring engines churning through the frigid night air, the plane which felt so huge when we were trying to get it under control on the Goose Bay runway, suddenly seems shrunken, in our minds, to a fragile shell of flimsy aluminum, puny and insignificant, forlorn and lonely in this boundless space. We are like a tiny gnat, adrift and vulnerable, lost somewhere among the endless galaxies. It is a humbling, abandoned feeling as we confront our paltry and inconsequential place in this limitless universe.

But we are seen and blessed by the Unseen, and after some hours, the signal from the Thule TACAN (tactical air navigation) station begins to stir on our instrument panel. The directional needle flickers and steadies. The little mileage window on the gauge commences its "Distance to Station" countdown.

We raise Thule Approach on the VHF radio, and the weather at Thule is clear, cold, and calm. Our anxiety begins to ease. In an hour, we are on the ground (it is 3 am, of course), precious cargo delivered. Out of crew duty time, we head for a well-earned rest at the BOQ. It has been a 30-hour day for us, and not without its "moments." Exhausted, we fall into our bunks and are quickly asleep.

IV. Blizzard, BMEWS, and the Base

Our orders call for crew rest at Thule, then return directly home with a refueling stop at Goose Bay. Had this been a non-critical mission, we would not stay over at Thule. The winter weather is too severe and unpredictable. Instead, we would have crew-rested at Goose, flown up to Thule, discharged our cargo, refueled, and "gotten out of Dodge" back to Goose. But here we are, and there is a price to pay.

The first impediment to departure roars in while we sleep. Fierce winds and blowing snow reduce visibility to zero, with the chill factor well below that. We are roused by the howl of the storm around "mid-day." Outside, it is black as pitch, the building is shaking and shuddering from the onslaught of the blizzard, and the temperature out there is about -50. It goes on like this for two days. We can't leave our quarters. We would freeze solid in about ten minutes. Besides, we wouldn't be able to see six inches—sure to become disoriented, lost, and deep-frozen, not 100 feet from safety.

Luckily there is a snack bar in our quarters (emphasis on "snack" and emphasis on "bar"). That, and a rack of dated magazines, are the sum-total of our entertainment options. Then, of course, there is always sleep, which comes naturally after too long in the "bar" part of the snack bar.

On the third day, the weather clears, and we get ready to leave again, only to find that the wind, snow, and deep freeze have done a number on our aircraft. The ground crew had lashed it down immediately after we landed, with no hangar large enough to protect it from the elements. The thaw-out and fixes will take several days.

During this period of the Cold War—the first half of the sixties—all our C-130 aircrews have Secret clearances. We need these because our missions—like this one—often involve clandestine work. So, the base commander, in gratitude for us getting him the critical parts he needs, takes pity on our marooned crew and arranges a tour of the secret BMEWS site.

We board our bus and, in the dark, crunching our way slowly over the newly plowed road, winding out across the tundra for about 12 miles. On the way, our tour guide, the base exec officer, fills us in on the history and purpose of BMEWS.

"There are two BMEWS sites–this one and one in Clear, Alaska. They were constructed starting in 1958 when it became apparent that ballistic missiles, tipped with nuclear warheads, launched from the Soviet Union would become an existential threat to the Free World. This site here at Thule was built with the cooperation of Denmark and Canada. It is much closer to the USSR than the one at Clear. We are 500 miles farther north than the northernmost point in Alaska. If the Ruskies fire missiles, we'll see them first here. That is why the parts you brought are so critical. We can't afford to be blindsided.

"We have two types of antennas. The first are big fixed grids that send out fan-shaped microwave beams, one above the other. The low fan allows us to see the missiles shortly after launch. The upper fan lets us judge the speed and general direction.

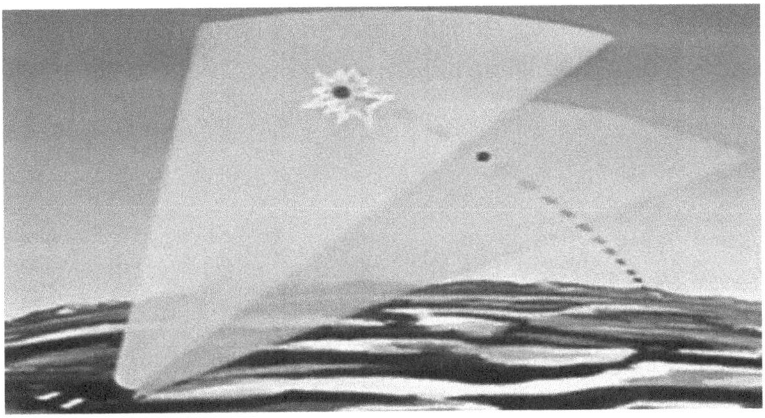

"Then the second radar comes into play. This one allows us to lock onto and track the target. It is a huge, hydraulically-driven parabolic reflector. It is protected from the elements by a standard radome, like those you have seen many times—except this one is 'just a tad bigger' than the ones you are used to seeing. The radome is 250 feet in diameter. The antenna is in constant motion, searching back and forth, and when the hydraulics lock it onto a target, the ground shakes."

We come around a bend, and a fantastic sight meets our eyes on the crest of a hill. Illuminated from the gloom by a vast array of floodlights, the BMEWS installation towers hundreds of feet above the barren tundra. It sprawls for nearly a mile along the brow of the rise. The gigantic "fish net" parabolic antennas—much bigger than football fields standing on their sides—range along the hill, dwarfing the 250-foot-diameter search dome.

Our host explains that the job of each of the "fishnet" monsters is to cover a specific sector of space, looking toward the USSR for any sign of a missile launch. Each covers a different sector, so together, they will see anything coming our way.

"The parts you brought are for the electronics in one of these transceivers," he explains. "They are doing the repairs as I speak. Since that one is currently not operating, I'll be able to show you how the tiny signal picked up by that antenna is amplified into a blip we can decode and display on the radar scopes at NORAD headquarters, the North American Defense Command, back in the States."

We come tumbling out of the bus next to one of the blockhouses that sit at the base of each of the gargantuan grid structures. It is calm now, but the cold is fierce. We are glad to get inside and slam the door against the icy air.

"This is one of the 'transmitter' buildings. See these ten-foot-high vacuum tubes"?

We gawk.

"They give off enough heat to warm the building."

"Wow! And in this weather"!

"They're part of the amplification circuitry. But a good part of the amplification comes from the waveguides. These work like the horn on an old Victrola. They pick up the small-signal wave from the antenna. The wave is projected through a big tunnel in the ground. The tunnel gets bigger as it gets further from the antenna, and this amplifies the signal. We gather that amplified signal and convert it through the electronics into speed, height, and distance information that we transmit to NORAD."

Wow again.

A BMEWS waveguide under construction

From the transmitter building, we go below ground to the inactive waveguide and take a walk-through. It is like a long mine tunnel, getting smaller as we approach the massive grid antenna that feeds it.

That concludes our tour, and back we go to the base to await our aircraft repair. At least the weather has improved to where we can get around the base in our cold-weather gear. It isn't that large, but it has a full complement of facilities—An Officers' Club, a base exchange, a commissary, a gym, a bowling alley, and a movie theater.

We establish a daily routine: Arise late, leisurely breakfast at the Officers' Club, over to the base exchange to pick up some more trinkets for the home front, on to the commissary for a few supplies for back in the room, then back to the "O" Club for a leisurely lunch. Off to the bowling alley to roll a few frames, then to the movie theater until it is time for happy hour at the club—beer a dime, martinis or Manhattans, a quarter. Lucky if we make it to supper in the dining room, guaranteeing another late and bleary arousal the next day. Remember, all this is done in the pitch dark. How do people survive being stationed here, especially in winter?

We have discussions at the bar with some "locals." The assignment here is for one year, unaccompanied by family. How do they survive here? Part of the answer is that they work; they believe their work protects the Free World and are dedicated to it. And, they assure us, the summer is glorious. So they can roam around the nearby "countryside," and explore the interesting terrain, flora, and fauna in the land of the midnight sun. And the temperature may even reach 45 or 50 degrees—above zero!

V. Heading Home

Every mission has two objectives. The first, of course, is *Complete the assigned task* and then, equally important to us aircrews, anyway, *Get everyone home safely.* We had succeeded at the first. However, the second seems to have become a challenge.

But on the seventh day, our airplane is ready, and the weather is cooperating, so we hurry to the flight line to "kick the tires and light the fires" before the meteorologist changes his mind.

We wave a fond farewell to Thule and take off into the faint glow on the horizon. As we approach cruise altitude, we can see the sun. Hurray!

But the weather is not as benign at Goose. Snow is squalling, but at least, this time, the wind is down the runway. We waste no time refueling and "get the flock out of there" because the forecast is deteriorating.

But the homebound challenges continue. We are not an hour on our way when we lose the generator on the number two engine. This is a mandatory shut-down to avoid starting an engine fire, so we feather number two, call Harmon AFB a hundred or so miles in front of us, and request an emergency landing. No dice. Harmon is below minimums. We call Goose, requesting a return. Nope. Below minimums now, too. Loring in Maine? Ixnay. Hanscom AFB near Boston? No way. The maritime Nor'easter has everyone shut down.

Flying on three engines, we continue through the snow and icy precipitation all the way down the coast to McGuire in New Jersey, letting down through driving sleet and freezing rain. *We can always restart number two,* we think, *if we need her, in case we have trouble with another engine.* Fortunately, no extra predicaments, and we break through the heavy, sodden overcast about three hundred feet off the deck and make a normal three-engine landing, spraying a sloppy salvo of slush onto the emergency vehicles that follow us down the runway. Over the radio, we apologize for our inconsiderate behavior and thank them for their kind attention; darned glad their expertise is not required.

We taxi to our spot on the ramp and egress the aircraft. Looking up at the number two engine, we realize there is no way we could have restarted it. When we executed the shut-down procedure, it deactivated the inlet duct deicing. The engine inlet is packed with ice.

It takes three hours to thaw out the duct and change the generator, and then we are on our way, the last leg of a long and harrowing mission. In a couple of hours, we are home—and so glad to be here.

Mission accomplished.

Bomb Run to Ubon

By Charles F. Willard

Many of our Air Force C-130 missions during the Cold War era of the 1960s were classified "Secret." Usually, the Department of Defense wanted to keep our adversaries in the dark and off-balance about what we were doing. As a result, on many of our flights, even our flight crew didn't know what we were hauling or why. Sometimes we didn't even know "where" until we were three-quarters of the way "there." But this secret mission was different: We knew what and where but not why.
I would not find out the "why" until decades later when the mission and the intrigue surrounding it were declassified.

****

Our journey began with a two-and-a-half-hour hop from our "home drome" at Charleston Air Force Base, South Carolina, to McGuire AFB in New Jersey to pick up our cargo. McGuire is next to Fort Dix, a significant supply depot. This was when the war in Vietnam was ramping up, and weapons and equipment were pouring from factories in the Northeast into our bases in Southeast Asia. Our planes from Charleston transported a lot of that gear.

When we got to McGuire, we learned that our cargo would be a belly-full of 500-lb. bombs. This was strange. Bombs were usually transported across the Pacific in large cargo ships that could carry them in quantities vast enough to keep ahead of their ever-growing use rate in that accelerating war. They were dropping thousands per day.

Now, a C-130 is a pretty big airplane. Its cargo floor measures 10 feet wide by over 40 feet long, about like two bowling alleys, side-by-side. It can carry sizeable military equipment like trucks and tanks. Still, it is only big enough for 36 of those 500-pounders. That's 9 tons of bombs, but less than half a load for a B-52 Stratofortress, one of the planes they were using to bomb North Vietnam. The BUFF, as she was called—Big Ugly Fat Fella (fella being a euphemism for a less polite word beginning with the same letter)—could carry 84 of those lethal 500 lb. blasters. *So* why *were we carrying this puny number of bombs halfway around the world when a ship could lug so many more?* Maybe, as the lawyers say, "time was of the essence."

Loading and securing the 36 bombs took a while. First, each was wheeled into the cargo bay on a specially designed forklift-like device, which would carefully position each bomb in the desired location on the cargo floor. Then it was lashed tightly into place with heavy chains so there would be no movement during the turbulence we would surely encounter during the long, cross-country, and trans-Pacific flight.

Up and down 36 times. The forklift chugged in and out, and the loading crew laboriously lashed each projectile into place. When they were done, a crisscross of chains laced the cargo floor. It would be a hazardous walk from the cockpit to the "porta-potty" in the rear. Especially if, as the lawyers say, "time was of the essence."

As the loading process dragged along, I had plenty of time to examine the bombs. At first, they looked like the standard 500-pound variety I was familiar with. But, as I looked closer, I saw a couple of curious differences. First, I noted that the fins at the rear of each bomb were hinged, like the rudder on our plane, and there were thin control wires attached to the hinged parts. These ran from the fins to the tailpiece, where they disappeared into the main body of the bomb. *Was this a steering mechanism?*

The military didn't have guided bombs in those days. As far as I knew, all they had were dumb old bombs. The bombardier toggled them, and they free-fell to wherever gravity and the winds would take them. Looking at our cargo, I wondered, *were these some new-fangled explosives with a guidance system? Could something steer those fins into a specific target?* I didn't know of any capability we had that could do that.

A six-sided plug, like a big hex nut, was screwed into the nose of each bomb where the fuse would go. That was the second anomaly: These plugs were much larger in diameter than those in a standard bomb. *Was this bigger space needed because something more would be housed there than just the standard fuse? Maybe some sort of control system?*

It would be many years before I pieced together the details.

From McGuire, the flight to our destination was long and, for a C-130 mission, routine. Our endpoint was Ubon Royal Thai Air Base near the border with Laos, about 35 flying hours and several refueling/crew rest stops away.

When we arrived, the "Follow Me" truck led us to a remote corner of the field. I expected to be met by an unloading crew with the same sophisticated equipment used at McGuire to load our cargo—a special forklift to make sure each bomb got safely off the aircraft and deposited into storage.

On the contrary, we were greeted on the tarmac by a rag-tag company of about a dozen Thai recruits—little Thai guys; I don't think any were over five feet tall. Little skinny men, all sinew, no fat, standing haphazard and anxious in what passed for formation. As I descended the steps from our aircraft, their sergeant approached, saluted, and explained in broken English that his men would now relieve the airplane of its cargo.

Proceed they did. The sergeant gave the order, and they broke ranks, all running toward the rear ramp to see who could be first into the back of the plane. Chains were flying off the shackled bombs, and the little Thai guys were rolling the cylinders down the ramp. Soon, bombs were careening everywhere in an uncontrolled stampede of men and steel. Several found their way under the airplane, bumping against each other and into the giant tires of the main landing gear.

Dodging bombs and men, I sprinted to the sergeant and had him call a halt to the chaos. I explained the best I could that these were special bombs and pointed to the delicate-looking steering mechanisms on the fins. Then I told him that I wanted the bombs removed one at a time in a controlled fashion, two men on each end of a bomb so that each could be restrained on its trip down the ramp and off to its destination in the nearby hangar. This explanation took about ten minutes of language barrier time and was accompanied by a much-animated flurry of gesticulations before it fully sank in. Experience playing "Charades" came in handy.

In due course, the cargo was safely unloaded—no blown-up airplane, no bent bombs, and we went on our way: another 35 flying hours and back home in one piece.

I hadn't thought about that mission for years. It was one of many that I flew into Southeast Asia during those Vietnam War years, but I remembered the bombs and the little Thai guys when I came across an article—I think it was in Air and Space Magazine—sometime in the 1990s. The Cold War was over, then. The Berlin Wall had fallen, and the Soviet Union had dissolved. Much of what went on during the Cold War years was being declassified. To my surprise, this story revealed that a second war—a secret war—was being fought then in Southeast Asia, one that grew to become the largest clandestine operation that the CIA ever ran. Here is the gist:

In 1962, as a part of the Geneva Accords, we, the U.S., had agreed with North Vietnam to guarantee the neutrality of Laos. The Accords stipulated that all foreign military forces were to leave that country. The U.S. complied—but North Vietnam did not. Instead, the Viet Cong continued to infiltrate Laotian territory and smuggle vast amounts of weapons and men into South Vietnam through jungle routes along the long, shared border.

Laotian Prime Minister Prince Souvanna Phoumo asked the United States for military assistance to stop this violation of Laos territory. This presented President Kennedy with a problem: How could he comply with Prince Phoumo's request without violating the Geneva Accords? If the United States got involved publicly and directly, it might trigger a response from China and the Soviet Union that could lead to World War III (Sound familiar?).

Here is the plan they put in place: The USAF and the CIA designed a set of covert operations to interdict the enemy supply routes. Spotter planes owned by the CIA airline, Air America, and flown by civilian "volunteers" would fly low over the Laos border and give targeting instructions to Thai, Laotian, and Hmong fighter pilots. The targets were the Viet Cong's weapons and ammunition depots, some hidden in caves. These targets required weapons with greater precision than any currently in the military's inventory. Scrambling for a solution, a few hundred prototypes of experimental "guided bombs" were kludged together and rushed from the laboratory, specifically for trial use on those missions.

Developed under the utmost secrecy, these were flown into Thailand by United States Air Force cargo planes. It turns out that this "secret war" was conducted along the Laotian border for several years, initially waged by a small group of daring and dedicated pilots who flew rickety, slow-moving, outmoded aircraft at jungle-top levels, braving intense ground fire so they could identify and destroy their targets. Ironically, we had brought them a brand-new weapon that could be used from their old airplanes—a new bomb that could be guided directly to its target. Reports indicate they used them to good effect.

A couple of examples: Spotter planes uncovered a suspected fuel depot in a cave. "...the fighter-bomber dropped a guided bomb on the cave entrance. The resulting explosion created a fireball 1,000 feet across, so hot that a passing raincloud was sucked into its vortex." And, "...I uncovered pallet after pallet of 122mm rockets. The secondary explosions went on for two solid days...."

What a revelation. My guess is—it's only a guess, but it seems reasonable—that the "secret war" was an ideal testing ground for these experimental munitions. Within ten years, PGMs (precision-guided munitions) had become common, with over 10,000 laser-guided bombs used in combat in 1973.* Targeting accuracy went from about a quarter-mile with "dumb bombs" to about ten feet with smart ones. What a savings in both munitions and collateral damage.

Who knew? Many years later, it seems that our Air Force C-130 transport crew had a small part in advancing the cause of a "secret war" and enabling the testing of a secret bomb which led the way to what now is a whole inventory of modern "smart" weapons.

**Congressional Research Service Report # R45996, June 11, 2021*

Typhoon

By Charles F. Willard

The Pacific–the largest ocean on Earth. At 64 million square miles, it is 16 times bigger than the United States. Its northwest quadrant lies west of the International Dateline and north of the Equator. The Marshal Islands, Caroline Islands, and the Marianas sprawl across it for thousands of miles, and just east of Guam lurks the deepest point on Earth, the Mariana Trench, seven miles below the sea's surface.

This single area of the Pacific spawns one-third of all the cyclonic storms on the planet. In the U.S., we know them as hurricanes, but there, they call them typhoons, and they strike fear into the hearts of the sailors and airmen who navigate across that area of the world.

I encountered them once in a while, back in the first half of the 1960s when I was piloting Air Force C-130 Hercules aircraft back and forth across the Pacific on the way to and from Southeast Asia. None of those encounters was pleasant, but this one will never leave my mind....

****

I look in on my sleeping son, fifteen months old now, kiss his cheek, tuck the blanket around him, and leave the door ajar. Then I hit the sack myself, easing into bed beside my exhausted, already sleeping wife. I try not to disturb her. She has been chasing after the boy all day, trying to keep the hyperactive little racer out of trouble.

We are looking forward to a rare good night's sleep, and I drift off in about a minute.

The bedside phone jangles me awake.

"Lt. Willard, this is Captain Owens, Charleston Air Force Base Operations. You are alerted for an 0400 briefing and an 0600 take-off. Pack for at least a two-week trip."

This is a shock. I just returned from a grueling two-weeker and have yet to catch up with things at home.

"Mmfff. WheremIgoin'"?

"Your destination is classified. You'll find out more at the briefing."

I glance at the numbers on the bedside clock radio: 3:00 a.m.

My wife is awake now. "What is it, hon?"

"I gotta go. Don't know where. Classified. They said, 'Pack for two weeks.'"

"Oh, no. Not again. Can I get you anything?"

"No. Go back to sleep. I'll drive the car to the base. You can get one of the neighbors to take you in in the morning. It'll be in front of Base Ops with the keys under the floormat."

"'K. Be safe. Love you."

"Love you too." I bend down, lay my hand on her cheek, and kiss her. "Bye."

I don't have to pack. My bulging B-4 bag sits ready to go in the closet, stuffed with assorted clothing, uniforms, flight suits, and gear. Before I leave the house, I look in on my boy. I gaze at him, the nightlight casting soft shadows across his small, blanketed form, watching him breathe for a moment. I am still touched by the miracle of this small, vibrant life and still in awe of being a parent....

It is time to go to work. Not wanting to wake him, I blow a silent kiss his way and head out into the night.

I am on the familiar road to the base, driving through the almost nightly late Spring Charleston drizzle. It is 3:30 in the morning. I don't want to do this anymore. Don't want to leave my young family alone for most of every month. Don't want to miss these precious days when my boy is learning and changing so fast. As I go, I reach the decision I've been wrestling with for some time. I'll share it with my wife when I finally get back home.

The date is 9 June 1965, a Wednesday.

****

The crew gathers at Base Ops for the 4 a.m. briefing, all of us dragged from our beds by similar phone calls. I glance at the faces. I have flown with most of these guys before. There are eight of us, an "augmented" crew: three pilots, two navigators, a flight engineer, and two loadmasters. No old majors or light colonels among us five officers; we are all lieutenants. No tech or master sergeant stripes grace the enlisted men's sleeves. Each of us is a low-ranking guy in his twenties. I might be the oldest at 26.

This is the way the Air Force does things. They train you well for your job, then trust you to take multi-million-dollar airplanes and precious cargos into any weather or situation anywhere in the world. They trust you to survive the experience and complete whatever mission they give you. They seem to have faith that their training instills skill, maturity, judgment, and mental toughness.

I wonder, briefly, if this is hubris or based on experience. Then I recall that in World War II, plenty of 19 and 20-year-olds were pilots and aircrew members, and they did their sometimes-impossible jobs with brave and competent honor. Compared to them, I guess we are old guys.

At the briefing, we find out nothing except our first destination. We will fly all the way across the country from Charleston, South Carolina, to Travis Air Force Base, just north of San Francisco, to pick up our cargo. Final objective: undisclosed.

"Travis?" we ask, "They have a transport squadron out there. Why don't they take the mission instead of making us fly eight hours cross-country to pick up a cargo that's already at their base?"

"No Travis aircraft or crews are available." is the answer. So what's going on here? Are we about to become a small part of some larger clandestine undertaking?

Out into the soggy-warm South Carolina night we go, to pre-flight the aircraft, crawling around in the dark with our probing flashlights.

Meanwhile, our flight engineer has climbed up to the flight deck to check out the aircraft systems: fuel, hydraulics, pneumatics, environmental, and electricals. He finds three discrepancies that need fixing before we take this bird across the country (and probably across the Pacific to some unknown—most likely godforsaken—destination).

The repairs take over three hours, so our 0600 take-off is rescheduled for 1000 (that's "ten hundred" in military jargon, in other words, 10:00 a.m.). At least it will be daylight when we leave, but this burns four hours of our crew duty time just sitting around waiting, drinking coffee, and speculating on our mission. What kind of cargo? Where are they sending us? How long will we be gone? Why all the secrecy?

We only know that our first stop is Travis. "Augmented crew" means we have 30 hours of crew duty time, not the usual 12-hour limit that a regular two-pilot, one-navigator crew would have. So that makes us think that we'll be stopping only briefly at Travis to load the cargo and refuel, then head straight out across the Pacific.

Eight guys in the dark about what is coming down. We know that the situation in Viet Nam is unstable, and "activity" is heating up. Are we headed there? Our final destination orders are sealed in a pouch, "**SECRET. DO NOT OPEN UNTIL DIRECTED,**" stamped in red capitals across the front. Nothing about what, when, where, or how we will be "**DIRECTED.**"

Usually, when we head west across the ocean, the next stop is Hickam Air Force Base in Hawaii, eight droning C-130 hours from the California coast. The delay in our departure might alter the mission plan. We may run out of duty time and have to stop for crew rest at some place short of where the bigwigs have planned—like Hawaii. Our spirits, at low ebb from the tedious, anxious wait, notch up a tick. The bachelors on the crew, none of whom have yet been to Hawaii, are buoyed by the thought: bikini-clad wahines on the beach at Waikiki....

****

Things get fixed, finally, and we get going. It is a smooth, non-eventful flight across "fly-over country" to Travis. We shoot the "Vacaville One" instrument approach—just for practice; the weather is clear—and grease the landing, rolling up to hangar number 15, as directed by ground control and the "Follow Me" truck.

The officers—we three pilots and the two navigators—head for Base Ops to plan the next leg of the flight and learn what we can about our "secret" mission. The enlisted guys—the engineer and the two loadmasters—stay with the aircraft. The engineer has to see how long it will take to fix some electrical issues that cropped up on the cross-country flight. Then, he will supervise the maintenance and ensure the bird is good to go once everything is repaired.

The loadmasters will see to the cargo. They need to understand the weight and balance situation, supervise the loading so that all is properly secured within the center of gravity limits of the airplane, and understand any special requirements for its protection in flight and during off-load.

At Base Ops, we are greeted by a Lt. Col. Darnell, who reveals that our crew rest layover is supposed to be at Wake Island, with a brief stop at Hickam to refuel. I do a quick calculation. We are already over halfway through our 30-hour crew duty limit, and Wake is about 16 flying hours away, not including the hour-or-more stop in Hawaii. We can't make Wake Island before we run out of time. "Uh, Col., I guess we'll have to crew-rest at Hickam." Shucks. Drat the luck. Our two navigators go off to Flight Planning to lay out the next leg, leaving us three pilots alone with the Col.

We ask about our final destination. "Sorry, it's classified. You'll find out more when you get to Wake." And what is our cargo? "That's classified also. It is wrapped up tight, and the pallets are sealed. So it's best if you don't speculate. But you do have five passengers."

"Passengers? They're military, I assume. Could I see the manifest?"

There are five names but no rank, serial numbers, or further information.

"Who are they? What do they do?"

"Not for you to know."

We all have secret clearances, yet they tell us nothing—curiouser and curiouser. So we meet up with our navigators, check the weather, grab a bite, and commandeer some box lunches to carry out to the rest of the crew.

I find the maintenance chief. "How're the repairs going?"

"The circuit breaker keeps popping on the number 2 HF radio. Do you want to go with just the number 1?"

"Not a chance. We've got a 2,000-mile over-ocean flight ahead of us to Hickam, and HF is the only way we can communicate when we get out of sight of land. So we need a backup."

"Roger. We'll keep working on it."

The sun is sinking toward the Pacific by the time they get the radio repaired.

All this time, the Travis freight crew and our loadmasters have been working on getting our cargo prepared for the flight. The whole process is being made more complicated by our five passengers. They are dressed in starched, pressed, combat fatigues, their khaki britches bloused into the tops of spit-shined combat boots *(What are these guys? Special Forces? CIA?)*, but no rank or service designation shows anywhere on their uniforms. A little older than ourselves. Muscular, tough, no-nonsense demeanor. Next to them, we, in our rumpled, sweat-stained, onesie nylon flight suits, look like a bunch of young ragamuffins. There are no smiles, no pleasantries. They are all business. They have a lot to say about how "their cargo" is treated and loaded.

There are four large, differently shaped masses. Looking at the crates beforehand in the hangar, I can't envision how we will fit them all in and secure them properly. In addition, there can be no shifting around during the turbulence we are bound to encounter. Then there is the problem of keeping everything within the CG (center of gravity) limits. If the load is out of limits, it makes us either tail-heavy or nose-heavy. Either of these conditions will put control of the aircraft at risk. That's why we have "loadmasters." The name is appropriate because they are experts—masters—at figuring all this stuff out.

We have to explain these harsh realities —with some firmness—to the "starched/pressed" guys. They can advise, but our loadmasters have the last call. The safety and survival of our aircraft, the cargo, the passengers, the crew, and the mission depend on the L/Ms doing their jobs.

When everything is loaded and secured, the belly of the plane is stuffed from the forward bulkhead all the way aft to where the rear ramp slants upward toward the ceiling of the cargo bay, under the tail. One container, long, flat-ish, and narrow, stretches nearly the length of the space and is lashed to the tops of the other three pallets. Getting that one to fit is the biggest challenge. There is no way to squeeze that container in once the others are loaded.

Finally, they decide that it must be loaded first, then raised to the roof of the cargo compartment and trussed to the ceiling while the other three massive items are placed and secured underneath it. Then it is lowered into position on top of the three and lashed in place. The base's carpentry shop has to build a special pallet to sit on the up-slanted tail ramp and support the rearmost part of this cumbersome, 40-foot-long crate.

There is barely room left for our five "guests." A C-130 cargo compartment is a torture chamber for passengers. The cavernous space can be alternately insufferably hot or blizzardly cold. The incessant, 104-decibel roar of the four massive turbo-prop engines, amplified through the thin aluminum skin of the fuselage, is debilitating—something like sitting on the inside of a rapidly beating bass drum. Ear protection is mandatory but ineffective. The wax earplugs dampen the din somewhat, but it is relentless and exhausting. By 21st Century mores, subjecting someone to this treatment for eight hours would probably be considered an immoral torture technique.

The passenger seats, web-backed, aluminum-framed canvas horrors, are stored against the sides of the cargo space. They are designed for temporary use by paratroopers before they gladly exit the aircraft over the drop zone. Anyone sitting in these sag-backed, bottom-numbing contraptions for more than an hour or two will need a month's treatment by a chiropractor to walk upright again.

Restroom facilities consist of a single, odiferous, porta-potty-type stool, which sits on a small fold-down platform suspended above the slanted ramp at the rear of the airplane. It has the luxury of a rumpled gray-green nylon curtain that can be unsnapped and pulled around for what passes as privacy. One's shoes show below the curtain hem to indicate occupancy. All the effluent deposits itself into the large plastic bag secured by clamps to the insides of the pot. The excretive essence lingers until the bag is removed after landing— another thankless loadmaster job.

Our senior L/M (senior because he is a staff sergeant, and his partner is a rank below that) unlimbers a set of the "torture seats" and directs our five guests to sit. He is the "stewardess" when we have passengers. (The term "flight attendant" doesn't come into use until later).

There is grumbling. "Don't you have room for us on the flight deck?"

"Sorry. We have an augmented crew. There is barely room for all of them up there."

Our L/M finishes briefing the passengers, shows them how to strap in tight, and retreats to his own torture seat in the rear. Loadmasters are the low men on the crew totem pole, but their job is as essential to a successful mission as any other crew member. Their habitat—and fiefdom—is the cargo space. They are the rulers of that domain.

It is 8 p.m., 2000 hours, by the time everything is sorted out and, we are finally ready to go. Welcome aboard the red-eye to Honolulu.

By this time, we are all dragging. On these long days stretching into night, we three pilots play musical chairs in the left seat. This is supposed to ensure that at least one of us is halfway alert enough to land the aircraft when we finally arrive wherever. Surprising how alert you become when you need to get that baby safely on the ground. Especially if the weather conditions are challenging. Adrenaline does wonderful things for alertness, exhausted or not.

We fire her up, get our clearance, and "put the wheels in the well" as the sun melts into the ocean. Our course takes us out past the Golden Gate, the lights on the bridge visible in the distance, blinking off to our left. The night weather is calm and clear over the eastern Pacific. It is a tedious, uneventful eight hours to Hickam, and we touch down there at 0230, Hawaiian time, and bed down in the barracks just before we run out of crew time. It is 8:30 a.m. back in Charleston. We have been on duty for 29 ½ hours. We pull the black-out curtains and hit the rack, exhausted. So much for the beach, the wahines, and their bikinis.

****

We are alerted 12 hours later, Thursday, June 10, at 1430 hours (2:30 p.m. to you civilians), to start our next 30-hour day—still no enlightenment about our final destination. At our afternoon breakfast, we probe our passengers, trying to find out what they know, but they ain't talkin'.

At Base Ops, they confirm that we are headed next for Wake Island, a tiny dot of an atoll just west of the International Dateline, another 8 hours away. There is nothing but ocean around it for hundreds of miles. A navigational needle in a vast, watery haystack. And no word about when we can open our secret orders.

Out at the airplane, we find that a valve in the HVAC system won't open properly. That won't do because the temperature in the cargo compartment depends on that valve working right—fifty degrees below zero is a little under the cabin comfort level. Fortunately, Hickam maintenance has the spare parts, but the repairs delay our departure until after dark. Fixes done and checked, we take off into the night and churn our way westward.

Before taking off, we check the latest reports for the weather between Hickam and Wake Island. There will be a high cloud cover most of the way. When there are few clouds above us, the navigator can find our way by "celestial navigation." He pokes his sextant through a port in the roof of the aircraft, finds known stars by their location in the heavens, and triangulates our position from this information.

But, if he cannot see the stars, the moon, or the sun, he will have to navigate by "dead reckoning." That means he has to estimate where we are by guessing our course and speed. These two are influenced by headwinds or tailwinds and how much we are pushed to one side or the other because of any crosswind. So he has to make his best guess at that, too. No navigational satellites or GPS in those days.

I had often wondered why they call it "dead reckoning." Sitting high over the bottomless ocean in the middle of the night, with the static from the HF radio crackling in my earphones, gauges all in the green, the engine-drone loud and steady, auto-pilot doing most of the work, my mind tends to wander a bit. *Dead reckoning: I guess it's because if we reckon wrong, we'll end up dead.* I think about Amelia Earhart and her navigator Fred Noonan, who never found Howland Island. Happy thought. My mind changes the subject.

I glance back at the navigator's station. He is hard at work. He has a pair of dividers in his hand, measuring a distance on his chart, then takes out his E6-B flight computer, moves the wheel around, stares at the result, and scratches his head like he's puzzled. *Uh-oh!* I think.

"Pilot to Nav. Having trouble?"

"I think we've got a lot more headwind than they predicted."

"I see a small break in the overcast up ahead. Maybe you can get a star shot to confirm."

"'K, thanks."

The hours go by. I have our ADF (automatic direction finding) radio tuned to the Wake Island frequency. It is a "line of sight" beacon, so we must be pretty close to our planned course, or we will miss the signal. From our altitude of 26,000 feet, we should be within "line of sight" at about 100 miles out. Sure enough, when our navigator calculates that we are within range, the wandering ADF needle begins to steady at about 10 degrees off our nose. So we bank left until the needle points straight ahead and follow the beam to Wake. It turns out the "dead reckoning" was almost "dead-on," and our "navi-guesser" has found the tiny dot in the middle of the endless ocean.

When we land at Wake to refuel, it is the day after tomorrow—time warp. Crossing the International Dateline, we have lost a whole day. We left Hawaii on Thursday, the 10th. When we land at Wake Island, the morning sun at our backs, it is Saturday, 12 June 1965.

"Cheer up," I tell the crew, "We'll get the day back when we finally reverse course for home and fly east, back across the Dateline."

Someone asks, "But what if we just keep going west and fly all the way around the world? We'll never get it back." I don't have an answer for that one.

While the groundcrew refuels our bird, we grab some grub at the canteen, then head for Base Ops. There we learn that our next stop is Andersen Air Force Base, Guam. It is five slogging C-130 hours farther west. That is all the info we get.

Our next burning interest is the weather. Typhoons have been active this year, crisscrossing the east-west air and sea routes, the area we will fly through. Several have made landfall in Guam, the Philippines, Taiwan, and Japan and done major damage. Of course, this is in the backs of our minds all along, as our westward, ho! track has squarely aimed us at "Typhoon Alley."

We hike over to the meteorology desk for the weather briefing. There is some storm activity.

"Yesterday, a tropical depression formed several hundred miles to our southwest. Ship and aircraft reports indicate she is headed west and is growing in size and intensity. Her wide-flowing 'skirts' will make for plenty of cloud cover along your route, and her counter-clockwise rotation will give you a tailwind to Guam and probably some moderate turbulence.

"She is not nasty enough to call her a typhoon yet, but watch her closely wherever you go over the next week or so. Right now, we're predicting she'll make typhoon status within the next couple of days as she gathers strength over the open ocean. We're guessing her track will be to the west for a while; then she'll brush the northernmost Philippines, turn north to Taiwan, then northeastward toward Okinawa before she peters out. She's gonna be a real bitch."

It's time to "kick the tires and light the fires," but we find another maintenance issue when we get out to the bird. The generator on the number four engine won't come up to voltage. Wake has a small supply of spare parts for C-130s, and, wonder of wonders; they have the relay we need. It takes all afternoon to diagnose and repair, and we finally take off into a hot, hazy evening sky, a mass of gray cloud build-ups visible on the western skyline.

We climb out toward the beautiful, eerie, occluded horizon. The last gasp of a fiery sun is straining its way through the stratified cloud mass, casting first, shifting hues of gold, orange, and ochre, then pink and rose. As the sun sinks below the ocean and the light ebbs, these calming colors morph into darker, more ominous shades: dried-blood reds, lurid prune-purples, and as the last light leaves, transform as we watch to smeared, grimy browns and sooty, darkening grays.

"Hey, Nav, get your best fix now. It looks like we'll be in the soup for a while."

"Yep, I got a pretty good one off a freighter that's headed west. They got a good sun shot about 20 minutes ago, corresponding to the last one I got. Moon's coming up behind us, too."

An opaque wall of cloud towers above us as we struggle for enough altitude to get over it. No dice. We enter the murky mass, full dark envelopes us, and the turbulence begins.

"Pilot to loadmaster, have your passengers cinch down tight and make sure there's no slack in the cargo tie-downs. We're in for a bit of a bumpy ride. Better get those guys some barf bags."

"Roger, Lieutenant. Already done. I think a couple of them are gonna need 'em."

We outrun the core of the gathering storm, but it jerks us around the sky a bit. With the tailwind, we make Guam in 4 hours and 50 minutes. The loadmaster has several full "puke parcels" to dispose of. Sorry for the buckboard ride, guys.

It is 2300 hours local time (11:00 p.m.) when we break below the overcast and see the welcome landing lights at Andersen. Our passengers, looking a little less spiffy than when they got aboard at Travis, are met by a black van and taken away separately. A standard blue

Air Force crew bus pulls up to take the rest of us to chow. A major with "Duty Officer" on his badge steps off to greet us. Maybe now we'll learn where we're going—no such luck. The duty officer gives us his spiel.

"Gentlemen, you will crew-rest here at Andersen. Alert for your next leg will be in 12 hours. Report to Base Operations at noon tomorrow. We'll have more information then. Sleep well. You are going to need the rest."

So, the tension continues to build. Sleep well. Right. A fitful night is ahead.

****

1200 Hours, Sunday, 13 June 1965. Andersen Air Force Base, Guam:

Surprise. When we get to Base Ops, Colonel Gray, the Base Operations Officer, meets us. On a Sunday, yet! We must be important. He ushers us five officers into a small, windowless conference room.

"Gentlemen, your sealed orders, please."

I produce the pouch, and the colonel breaks the seal.

"You may remove your orders."

I drag the sheaf of papers from the envelope. Here is the operative portion of the top sheet. It is typewritten and mimeographed in purple ink, as all our orders were in those days. (Copiers were in their infancy and unheard of in the USAF hinterlands).

41ST AIR TRANSPORT SQUADRON (M) AAM 1443/160
1608TH AIR TRANSPORT WING (H) (MATS)
UNITED STATES AIR FORCE
Charleston Air Force Base, South Carolina

FLIGHT ORDER)
628)

9 June 1965

The following crew members (Security Clearance SECRET) will proceed on or about 09 June 65 in such aircraft as directed, from Charleston AFB, South Carolina to TanSon Nhut, Viet Nam to perform an airlift mission, and upon completion will return to proper station on or about 09 Jul 1965. Variations in itinerary authorized. Government quarters and messing facilities will be used. Travel via military aircraft is directed when available. Authority: AFM 60-1, AFM 10-3, MM 55-1. Fund Citation: 57X4922.030 685-6508 92318 2113 2133 2153 S528300. 57X4922.030 686-6508 92318 213300 S528300.

So. We are headed to Viet Nam.

"Colonel, do you know what we are supposed to do when we reach Tan Son Nhut?"

"You will offload your cargo and await further orders."

"It says, 'return to proper station on or about 9 Jul 1965.' Does that mean we'll be gone for a whole month? We were told to pack for two weeks."

"That is a contingency. You could be gone for two weeks, or it could be two years. You are part of a larger operation. We have a deteriorating situation in Saigon that we hope can be concluded with a quick, positive outcome. Your passengers and cargo may be able to play a part in the successful resolution of a dangerous situation. That is all I can say."

I feel a chill go down my spine. Deteriorating situation. Larger operation. Two weeks to two years. What have we gotten ourselves into?

Off to Flight Planning. At least the tropical storm we skirted on the way in is still behind us. The weather guy says it is gonna be a big, mean one, and they are bracing for its arrival. But for now, we have outrun it. We lay our course westward to Saigon, Viet Nam, Tan Son Nhut Air Base. Our flight will take us over the Philippines' Luzon Island, south of Manila, then across the South China Sea into South Viet Nam. Another eight-hour flight, and it should be clear sailing, weather-wise.

Tanked full of fuel, we get into the air on schedule with no maintenance issues. As we bank around toward the west, we see the angry-looking storm glowering way off on the eastern horizon. Good riddance. About halfway into our flight, over Luzon, we give our hourly position report to Clark Air Force Base Control.

"Clark Control, MATS 37787 at 28,000 feet."

"Roger, 787. Read you five by five. Stand by for encoded message."

Encoded Message? What the bleep is this about? I think to myself as I get my pencil ready.

"MATS 37787, Here is your message: You are to proceed from your current position directly to Romeo Alpha Foxtrot X-ray Lima Delta. You have clearance to cross Bravo Zulu Tango Kilo. At your destination, you will report to Flight Operations Officer. Read back."

I read back the clearance, then scramble for the code book.

"Egads, we're headed for Bangkok! Don Muang Royal Thai Air Force Base. And we are cleared to overfly Cambodia. I better double-check this."

I get back on the radio: "Clark Control, MATS 37787. Confirm we are cleared to cross Bravo Zulu Tango Kilo."

"MATS 787, Clark. That is correct."

"Roger. 787 out."

On the intercom now: "Pilot to Navigator, did you copy all that? What's our course into Bangkok?"

"Gimme a minute.... Turn right to 265 degrees. We'll cross into Viet Nam airspace over Qui Nhon. That'll keep us south of Laos."

"Roger." I dial in the new heading. "Loadmaster, tell our passengers there's been a change of plan. We've been rerouted to Bangkok."

****

0100 Hours, Monday, 14 June 1965, Flight Operations, Don Muang Royal Thai Air Force Base, Thailand:

It is a bustling place. Here it is, the middle of the night, clerks' fingers are flying across typewriter keys, and orderlies are dashing between offices with sheaves and folders.

I dodge a couple of fast-moving bodies and report to Colonel O'Keefe of the USAF, the Director of Flight Ops.

"Lieutenant, your passengers and cargo will be off-loaded here. The current situation is chaotic. We've got a few balls in the air at the moment. We don't know what is next for you or where you will be going. It will take some time to sort this all out. In the meantime, get a good night's sleep. You and your crew are granted 24 hours of crew rest. So, if you want, you should have time for some sightseeing and shopping in Bangkok tomorrow afternoon. Report back here at 0100 hours Tuesday morning. We should know more then."

I am completely flummoxed now. Hush, hush, rush, rush, secret orders to Saigon, then rerouted in mid-flight to Bangkok, now suddenly a 24-hour break in the action. And again, no info on what's going on or what's coming next. *Thanks a lot,* I think. *You call, we haul. That's all.*

No use worrying about it. We get a decent night's sleep, wake up late, and have a hearty brunch at the base O-club. The five of us sit at a large round table. To the left of each chair stands a young Thai girl. They unfurl our linen napkins, place them on our laps, take our individual orders, fill and refill our coffee cups, anticipate our every need, and are our personal servants for the duration of the repast. Interesting custom. Don't try this at home.

****

Before we catch the bus for our afternoon in Bangkok, I go out to our airplane to check the progress on a couple of maintenance items we wrote up when we landed. They have towed us into a humongous hangar, out of the wet from the overnight rain. There, across from our aircraft, in the final stages of assembly, sits what must have been our cargo. My mouth drops open. It is the most evil-looking combat helicopter I have ever seen. Nothing like the pudgy, lumbering Hueys I am familiar with; it is small, sleek, and nasty looking. Twin Gatling guns protrude from under its narrow, rakish nose. Rocket cluster pods are cantilevered from the sides of the fuselage. All black. No external markings. Oversized twin jet engines nestle behind and above the tandem cockpit, bracing the rotor shaft. The scythe-like rotors span close to forty feet. They must be what was in the long, flat crate that sat atop the other three and barely fit within the length of our cargo bay. It has the other-worldly look of one of those astonishing, way-before-its-time creations conjured up in the Lockheed Skunkworks back in Burbank.

"Lieutenant, you don't belong here." It is the head honcho of our five passengers.

"I certainly do. I came to check on my airplane."

"Well, get about it then, and stop gawking at what is none of your business."

That is the end of the conversation. Not wanting a confrontation, I turn on my heel and go to look for the maintenance chief.

****

We find our way to the main gate and hop a jitney bus for town. The base is about 40 clicks (kilometers) north of the city. Our sightseeing mainly takes place from the windows of the bus, open to the hot, humid breeze. As we close in on Bangkok, the traffic gets heavier, a mix of animal carts, rickshaw-type conveyances, strange-looking autos of various sizes, and vintage "lorries" and rattletrap buses like ours.

Occasionally, we see plodding elephants, their curled trunks balancing long tree boles or other assorted burdens, their boy handlers at their sides to coax them this way or that. As we near the city, the golden spires of temples appear—we gawk, neophyte tourists not knowing or appreciating what we are seeing. Unfortunately, we have had no chance to bone up on the local tourist attractions. My only exposure to Thai/Siamese culture has been from seeing "The King and I," and I hardly think that "Anna and the King of Siam" accurately represents modern Thailand.

We wander around the captivating city for several hours, threading our way among the crowded streets and shops, enchanted by Bangkok's foreign customs, culture, and constructions. We do get some shopping in. My college roommate is getting married soon, and I need a wedding present. In a little shop on a side street, I find a unique and perfect gift: a set of brass picnicware with rosewood handles, complete with a velvet-lined, partitioned storage case, also rosewood. Getting foot-weary now, we find the restaurant recommended by a fellow officer we met at brunch and enjoy some good Thai beer and cuisine—the same "little Thai girl" serving routine. We could get used to this, but not recommended. Now, it's a scramble to return to the jitney station and grab the last bus back to the base. It is a snoozy ride as the soft, soggy night air pours through the glassless windows.

****

0100, Tuesday, 15 June 1965, Flight Operations, Don Muang Royal Thai Air Force Base, Thailand:

We report on schedule, somewhat bedraggled from our long afternoon and evening in Bangkok, and are met by the same Col. O'Keefe. We learn that the original crisis—the one that precipitated our orders—has eased.

Later, I pieced together what was going on then—the political/military situation during the days surrounding our operation.

Here's the scoop:

7 June 1965 (two days before we are alerted for our secret mission): General William Westmoreland, the U.S. commander in Southeast Asia, reports to the Joint Chiefs that the North Vietnamese invaders are stronger than ever, the South Vietnamese forces are taking heavy casualties, suffering low morale and high desertion rates and are unwilling to take the offensive. "I see no course of action open to us except to reinforce our efforts with additional US forces as rapidly as is practical."

8 June: President Lyndon B. Johnson authorizes General Westmoreland to commit American troops to combat missions in Viet Nam.

9 June: The Viet Cong (North Vietnamese guerrillas) overrun Dong Xoai, a district capital only 60 miles from Saigon (the capital of South Viet Nam). For a while, it is feared that Saigon (and with it, the government of South Viet Nam) is in danger of collapse. Instead, it becomes a bloody five-day battle in which over 400 South Vietnamese soldiers and 7 Americans are killed, with 12 missing. Many are wounded.

12 June: The president and prime minister of South Viet Nam resign. Military leaders take their place.

12-13 June: The enemy forces are turned back by U.S. airstrikes and massive reinforcements brought in by U.S. helicopters.

Were we to have some part in this do-or-die battle? Did that lethal-looking helicopter we transported have a specific death-dealing assignment it was supposed to accomplish—might still have to attempt?

We never find out. Such are the vagaries of the Cold War in the mid-nineteen sixties.

****

But now there is a new urgency. Col. O'Keefe gives us our orders. "Lieutenant, you and your crew have been assigned a "med-evac" mission. A major battle in South Viet Nam has resulted in heavy casualties. The hospitals in Saigon are not equipped to deal with some of these wounds. You will fly to Tan Son Nhut, on-load some of the injured and fly them to Kadena Air Base in Okinawa, where we have adequate facilities to treat them. Get a good weather briefing. There is a major typhoon somewhere near your route."

They have named her Dinah. She is the storm we skirted coming in from Wake, and she is now full-blown and headed right across our path toward Okinawa. I remember the words of the weather officer at Wake Island: "Wherever you're going over the next week or so, watch her closely...she's gonna be a real bitch." We'll have to fly right through her this time. With critically wounded soldiers on board who need the medical attention they can only get in Okinawa. We have no choice.

So, we get to Saigon after all. It's less than a two-hour flight. Along the way, our loadmasters rig us for med-evac. The C-130 can carry up to 64 litter patients with their attendants. The litters are stacked in two rows, up to five high, down the center of the cargo compartment, leaving room for the medics to work from each side. If flying back there is torture for

troops sitting upright, think what a loud and jarring ride it must be for wounded litter patients. And we're gonna fly 'em right through the middle of a category five typhoon. Lord, help them. And us too. It ain't gonna be no picnic.

Our stop at Tan Son Nhut takes two hours to refuel and get everyone loaded and as comfortable as possible. We explain that it will be a rough, unpleasant ride, but we will get them there safely. There are some terrible wounds—a few burn patients. I can't imagine the pain they are suffering and what this rodeo ride will do to them. They have tolerated the horrors of war and must now deal with its aftermath.

It is about 1,800 miles to Kadena by the best "avoidance" route available, about 6 hours in a C-130. Six hours of agony for the boys in the back. And terror, too. Even if you are hale and hearty, it is not fun to ride blind through a storm in the back of a bucking cargo aircraft, wondering if the pilot knows what he is doing, wondering if the infernal contraption will hold together, wondering if your eternal rest is going to be at the bottom of Davey Jones's Locker. The medics will sedate the ones in the most pain. The others will have to grimace and bear it.

We take off into early morning sunlight, headed northeast past Qui Nhon on the Viet Nam coast and out over the South China Sea. Our course leads us between Taiwan to our left and the Philippines to our right. Someplace between, we will pick a fight with Dinah.

Our climb is steady as the weather sours. Dinah looms as a massive malevolent presence directly to our front. I'm not sure how severe a shaking we will get or how brutal a wrenching we'll receive from the updrafts and downdrafts. It is a couple of hours of expectation and anxiety before we enter the gates of Valhalla.

With the autopilot still flying the aircraft and the gauges in the green, my mind wanders again. *How many parts does a C-130 have? Maybe it's millions. How many rivets? It's all held together with rivets. There must be a million of those! Every time we take off and climb to altitude, we pressurize the airplane. That puts all those rivets under stress. We do that over and over. When we fly into turbulence, it puts all those rivets under stress. What if they start popping?*

I have flown through storms but have not yet penetrated a typhoon. I'm not sure what to expect. The Coast Guard uses C-130s as hurricane hunters. They fly right through hurricanes. Typhoons are the same thing. They survive, so I guess we will. But have they ever flown through a Cat 5 storm? Certainly not with a belly full of wounded soldiers.

I look out at the wings and see those four big turboprops whirling, looking powerful and strong. *Look at those babies, spinning. Look at those big, sturdy wings—nothing to worry about.* Then I remember the Lockheed Electra, a passenger plane from a few years ago. Similar turboprop engines, wings constructed the same way, made out of machined aluminum slabs, no main spar, held together with rivets. Unfortunately, the engineers made a mistake, a miscalculation. When those planes flew into a storm, the turbulence created a resonance that tore the wing apart. There were three tragic crashes before they found and fixed the problem. *I hope they fixed it good—and designed this one better.*

We begin to pick up some turbulence. Just a slight flutter, the stuff we call "chop," and I watch the wings start to flex in response. *More flexing of those rivets,* I think, then there are the engines. *The jets spin at over 13,000 rpm. And each engine has a big, complicated gearbox. It gears down the 13,000-rpm engine speed to 1,000, so the prop tips won't break the sound barrier. Engines with super-hot turbine buckets flying around at hyper speed. The centrifugal force must be enormous. Gearboxes with all those straining gear teeth meshing and un-meshing, pulling the airplane through the air. The stresses must be terrific.*

I think about John Glenn in 1962, sitting atop an Atlas rocket, ready to launch: "How would you feel, sitting on top of 2 million parts, all built by the lowest bidder on a government contract?" Gulp.

Enough of this. The typhoon glowers above us—no way to get over or around. Nothing to do but plunge straight into her at what we hope is her least pugnacious area.

"Pilot to Loadmaster, we're beginning penetration. Make sure the caregivers are all strapped in and everything's secure."

I sound officious in my own headset. *Of course, he's got everything battened. Sounds like I'm implying that he doesn't know his job.* Nerves, I guess.

I say a little prayer for those poor guys back there and for the rest of us, too. We enter Dinah's outlying skirts, and the sun disappears, darker and darker as we go until it is night in the middle of the day. The clouds, thick and roiling, are jerking the big aircraft around pretty severely now. Whoops! Here she goes. I kick off the autopilot before it disengages itself.

It's like we just came out of chute number one on ole Dinah. She's mean and ornery, and she ain't never been ridden. We got to stay on her back somehow till the buzzer sounds. I got my spurs deep in her flanks now, a two G updraft followed by two G's down. The stomach hits the boots, then the roof of the mouth. Jerk one way, yank the other. Front hooves planted, rear end in the air, then reverse, then every which way at once. My feet on the rudders, both hands on the yoke, too busy trying to keep her right side up to be scared. That will come later. Then another colossal updraft. We gain 2,000 feet in seconds, then down. It feels like the wings will bend double. The bottom falls out, and the breakfast comes up, or rather, stays where it is, and I fall out from under it. I swallow it back down somehow before it spews itself over the gauges in front of me. I need to see those. They are my only reference to which way is up.

The vicious, incoherent violence settles itself into a more regular galloping, lurching ferocity, and then I see it.

Crackling on the windscreen, eerie blue-green arcs of mini-lightning flicker and dance across the outside of the NESA glass, a ballet of glow, glint, streak, and sparkle. I look out at the wings. A ghostly green aura dances along their leading edges, and at the tips of each whirling propeller hovers a blurred incandescent halo that shimmers as it spins.

Back in the cockpit, the surreal unfolds. A dim, luminous, grapefruit-sized, blue-purple orb of plasma forms from nothing. It bounce-rolls on a zig-zag path across the flight deck floor and disappears as quickly as it came. A split second. Did I really see it? For a moment, I doubt my own sanity. Then I recall, *I have heard of this: St. Elmo's fire*.

The turbulence abates to "moderate-to-severe." Then torrents of rain lash the windscreen, flooding up, down, and sideways. The ship feels more controllable now, but Dinah is far from exhausted. I, on the other hand, am pooped. The back of my flight suit is sopping with the sweat of exertion and probably nerves and fear. My arms and legs ache from the constant fight with the yoke and rudder pedals. I feel like I have been in the saddle of this bucking bronc for hours. But, it turns out, in twenty minutes, we break through into the sunshine.

I have had enough. I "put one hand around the other and let that pick-up man on in." That's all right. I stayed on her for a hell of a lot longer than the standard 8 seconds.

"Copilot, you have the aircraft."

I climb out of the left seat, stagger down the ladder to check on the souls in the back, and apologize for the bronco-busting rampage. The medics say they seem to come through it ok, except for a few full barf bags. I get some undeserved wan smiles and shaky thank-yous from those who are awake. I think: *The soldier's lot: fear, pain, and grit.*

We land at Kadena, and the ambulances meet us. Our whole crew goes down to the foot of the rear ramp to see our troops off the plane. We stand at attention and salute our brave soldiers. The things they suffer for the rest of us.

****

The rest of the mission is pretty tame. We get refueled and go on across the Dateline to Midway, a ten-hour flight. Sitting high over the ocean monitoring the gauges and radios while the autopilot flies, the navigator navigates, and the flight engineer runs the aircraft systems, I look at my hands. The hands that have, by this time, guided airplanes through take-offs and landings (and bronc-bustings) to distant places around the world, and I think, *I want to do something more with these hands; I want to build things.*

We crew-rest at Midway. A highlight is watching the goony birds (Laysan albatrosses) return to the island to nest, alighting on land for the first time after months at sea. They come down final approach just like an airplane: full flaps (wings gliding, adjusting to the breezes, fully deployed to their maximum 6-1/2 feet span) and gear down (large webbed feet extended, reaching for the ground). Only, after so long at sea, they forget the technique of a shore landing. Instead, they come in, expecting to water-ski along the surface while dissipating speed until they settle gently to rest.

It is a rude shock as they encounter the hard, gritty surface of an asphalt parking lot or road. The feet contact the ground, skid on the rough surface for a short, foot-scorching distance, then momentum takes over, and they go head-over-teakettles into a squawking, flailing upside-down pile of wings, feet, and feathers. They struggle to their feet, dazed and disgruntled, shaking themselves and quickly picking up one foot and then the other as if saying, "Ooh! Ooh! That stings! That burns!" They come in like that by the hundreds.

For us, then, it is Midway to Travis, to Kelly AFB in San Antonio, to MacDill in Florida, and finally home to Charleston on 19 June. Routine flights, routine cargo. Home in eleven days, not two months or two years.

LBJ, wanting both guns and butter for his Great Society programs and the escalating war in Viet Nam, is looking for cost savings wherever he can find them. To provide some of those savings, the Air Force decides it can make do with fewer pilots and offers an "early out," hoping to get rid of a few old majors who have hung around since World War II and Korea but have no hope of another promotion. Few have come forward, content to serve in that rank until retirement. In a pinch, the USAF extends the offer to all pilots.

Given this opportunity, I make my decision. Being away from home all the time in a risky career with an uncertain future is no way to raise a family. I have a wonderful 15-month-old son, and my devoted, loving, long-suffering wife is three months along with our second child. I am not being fair to her, and I have served my country for four years. Enough. I will be a family man now and an engineer and build things.

The Shackleton

By Charles F. Willard

When we landed at Chateauroux Air Base in central France late in the afternoon of 3 July, our Air Force C-130 crew had been flying for eight days with minimum rest. Our missions had taken us back and forth across Europe and the Middle East from Oslo to Tel Aviv and a dozen points in between. Finally, we were due for a break—a couple of days on the ground, uninterrupted by flying.

The respite came at the right time. We were exhausted, and tomorrow was the 4th of July. At that time, in the first half of the 1960s, the trauma of World War II was still fresh in European minds. Tomorrow there would be a commemoration of America's part in securing that victory. Our NATO allies had flown in planes and stunt teams for a holiday air show, and we could watch it. From the ground!

The "Follow Me" truck led us to our spot on the crowded ramp, where a wand-waving ground crewman motioned us into our parking position next to a strange-looking aircraft with Royal Air Force insignia.

It squatted there on the tarmac like an ugly gray toad. Its substantial twin-tailed rump hunkered low to the ground. The fuselage slanted upward from there, propped by its main gear, and protruded past the cockpit, eventually ending in a pig-like snout. My rural upbringing made me think of a snooty hog sniffing garbage with his nose in the air.

A sac-like goiter of a radome drooped under its reptilian chin, and bumps, bulges, blisters, and antennae of various configurations stuck out around its body—the toad had warts.

Four oversized engines jutted from beneath the long, slab-like wings. Later, I learned these were V-12 Rolls Royce Griffons, 2,450 horsepower each. They seemed out of place on this ungainly machine, looking better suited for the front end of a sleek, powerful fighter plane like the Royal Air Force's Supermarine Spitfire made famous in the World War II "Battle of Britain," or a North American P-51 Mustang, the scourge of the Luftwaffe. But here they were, adorning this hulking, crouching brute.

The oddness continued. At the front of each engine was a phallic hub sprouting not one but two propellers. By the differing pitch of the blades, I could tell that they counter-rotated! That is, each spun in the opposite direction from its mate. What a maintenance nightmare those complicated contraptions must present—gearboxes, seals, bearings, you name it. They looked like a quartet of giant roto-rooters, ready to puree me into a spray of red haze if I stepped too close.

I recognized this incongruous-looking plane. I had seen one from a distance when we were on the ground a week or so ago at Prestwick, Scotland. It was the Avro Shackleton, named for Ernest Shackleton, the famous British Antarctic explorer. The aircraft descended from the more glamorous, more handsome Lancaster, the UK's principal bomber in

World War II. The Shackleton was aptly named. It also had an explorer's mission—to hunt for enemy submarines.

Lugging our B-4 bags, our tired gaggle stumbled aboard the waiting crew bus and headed for quarters. We got cleaned up, then, tired but thirsty and famished, made a B-line for the clubs—separate ones for the enlisted men and the officers. No off-duty fraternizing, then.

The Officers' Club was mobbed with an international mix—French, Italians, Turks, Americans, Norwegians, Greeks, Germans, Brits—most of them here to fly and show off their aviating skills in tomorrow's big show. It was a loud and boisterous atmosphere, everyone competing, telling their own war stories—"There I was..."—in several different languages.

Mingling in the crowd, I found myself standing next to a Brit.

"Oi say, Mate, did you just come in on the Herc?" He used the nickname for our airplane, the C-130 Hercules.

"Yep. We've been buzzing around Europe and the Mid-east, repositioning troops and supplies for the last week or so."

"Saw you land. Oi envy you, Mate. You get to fly all over. Oi fly that "Shack" out there that you parked next to. We don't get to see much. Just orbit over the ocean lookin' for Ruskie pig boats, y'know? Oi'm Flight Leftenant Jack 'arrison."

"Lieutenant Chuck Willard. Pleased to meet you, Jack." We shook hands. He pumped mine as though we were long-lost friends. "How about a beer? I guess you Brits call it 'a pint'."

"Righty ho. Oi sees Oi've got a new best bud, Oi do."

I wondered if my American accent delighted him as much as his Cockney one charmed me. We hit it off right away, swapped stories, and ended up eating dinner together, still gabbing away about our flying experiences. During the evening's conversation, it developed that Jack had a problem. He was supposed to fly his "Shack" in tomorrow's air show—a low-level demonstration—but the Flight Surgeon had just grounded his copilot. Ear infection. "Blimey. Bugger all. Blast the luck."

"Gee, that's too bad, Jack. I'm sure you were looking forward to it."

"Yes, Oi surely was.... Say, Mate, Oi've got a bit of an idea 'ere. *You* could be me blasted copilot!"

"Now, wait a minute, Jack. Is that even legal? An American copilot on an RAF airplane?"

"Sure, it is. Me mates won't mind. Besides. Who's to know?"

"Yeah. Besides. I'm not checked out in a Shackleton. I don't even know what manifold pressure is. The Herk engines are turbos, not 'recips.'"

"Me engineer handles all that kit. Oi'll do the flyin'. All you have to do is handle the checklists and the radios and enjoy the ride."

So we sealed the deal with one more pint for a nightcap.

****

The next morning, waking to the cold light of dawn, I wasn't so sure about this. *What kind of deep do-do would I get myself into with the USAF? An unauthorized flight in a British airplane.* But then, I thought, *Who's to know?* Besides, I had just signed up for an "early out" from the Air Force. *What could they possibly do to me if they found out? Probably something, but what the heck? Better to ask forgiveness than permission.*

So, that afternoon I showed up at the Shackleton at the appointed time. Jack introduced me to the rest of his crew. "Righty ho! We'll show you a right good time, we will. Show you what RAF aviatin' is all about, we will!"

We were standing next to the nose of the Shackleton, and Jack was briefing us about what he would do in the demonstration. At that moment, I learned that this airshow wouldn't be like any other I had seen. A tiny Fiat Gnat, one of the aircraft flown by the Royal Air Force aerobatic team, the Red Arrows, buzzed our position—swooping in so low that he was using "ground effect" to keep from contacting the tarmac.

The little mite of an airplane (its wingspan was only 22 feet) streaked right between the Shackleton and our C-130. We were parked with only about 40 feet between our wingtips. Did he actually fly under the wing of our Herc? Perhaps he did! It was hard to tell from where I was standing. It happened in the blink of an eye.

I was astonished. Nothing like this was allowed in the American Air Force. Any pilot caught in such a stunt would be court-marshaled and dishonorably discharged for flagrant risk of lives and equipment.

But Jack and his crew seemed amused at the incident and appreciative of the skill it required. It made me wonder what Jack had in store for us.

We clambered up into the cockpit. The layout here was more cramped and less complicated than the spacious, complex C-130 flight deck. The instrumentation was more rudimentary. The sparse array of gauges was World War II-era, small, black and white, with no colors to help distinguish one from another during a rapid cross-check. Visibility from the pilot positions was good out the sides, but with the nose in the air, you couldn't see where you were going, looking straight ahead.

We had a half-hour before our scheduled block time. Jack used it to brief me on the copilot's duties and a few essential emergency procedures I would need to know if something unforeseen happened.

"So, Oi'll need yer on take-off and landing, but Oi can handle 'er meself in betwixt. So feel free to roam around the old crate and see what it feels like to lie in the bombardier's slot in the nose or the tail gunner's position in the tail. The old girl will give yer a few thrills, Oi wager."

"Thanks." *He'll probably try to scare the behoozes out of me with this low-level stuff,* I suspected.

We strapped in, and I called the tower, trying for my best English accent, "I say, Chateauroux tower, Whiskey Romeo 990, taxi-for-take-off instructions if you please."

Over in the left seat, Jack burst out laughing. "If that ain't the epitome of slaughterin' the Queen's English, Oi ain't never heard it, mate."

He should talk, I thought to myself.

Jack got us underway. It was tricky because there was no tailwheel steering, just differential throttles and braking to get her pointed right, and we couldn't see out the front. But the brakes were nearly useless for steering because they were pneumatic. You had to press the pedals and count slowly to three before anything happened. By that time, you'd be in the bushes.

When we got to the runway, crew members went back into the airplane's rear to pull lock-pins out of the flight controls. Then, ready to go, with the brakes locked, the engineer, not the pilots, managed the throttles, making sure each was precisely set so that we would go straight down the runway, not be pulled to one side by uneven thrust.

With power set at the max, the roaring whine overwhelmed us. The Griffons sounded like four giant vacuum cleaners with bad bearings. I looked out at those whirling, counter-rotating props and found that I was staring right into their spin path. If one of those babies cut loose, it would come right through the cockpit like a butcher slicing bologna. I had an instant urge to get the take-off over and get out of the line of fire.

Jack released the brakes, and we were off —sort of. It was as though she had to decide whether to bestir her 70,000 lb. self or not. Then, slowly, she began to roll. At 50 knots, Jack pushed the control yoke forward. This lifted the tail wheel, and we could finally see to the front. Then, around 100 knots, she flew herself clear of the ground. It took about 3,000 feet of the lengthy runway.

"Retract the undercarriage."

"Oh. I guess that means, "Gear up."

"Righty ho. Now, flaps to the flying position."

"At least those have the same name. Flaps coming up."

We climbed straight ahead, gathering speed until we were several miles from the runway, then Jack executed a one-eighty and descended to ground level. We came roaring back down the runway at max speed and power, the vacuum cleaners at banshee howl and the roto-rooters just feet off the ground.

We could see the crowd, close at hand in the grandstand, applauding as we roared by. Jack pulled us up into a tight max performance half-chandelle, and we exited the area on a course perpendicular to the runway. Having disappeared from the scene, the crowd would forget about us as they watched the next stunt group do their thing.

I excused myself from the right seat and made for the nose compartment. Lying on my belly at the bombardier's station, I had practically a worm's-eye view as we scooted over the landscape. I felt like I was riding on the nose of a bullet. Not that it made any difference if something were to go wrong, but, lying here, I would be the very first one in this irresistible force to meet with any immovable obstacle that might get in our way.

Jack was making an extended, diving turn, heading back toward the field. He had our props sweeping the ground again, so low I was looking up at the tops of the random trees and large haystacks that dotted the French farm fields. As we streaked back toward the airbase, I could see Jack was aiming us right at the crowd in the grandstand. *Geez, Jack, don't get target fixation!*

I was close enough to see wide-open mouths and people ducking for cover as we flashed over their heads, whirling props clearing the top row by what seemed to me to be only a couple of feet. I'm sure it was more than that, but it was an instant thrill, let me tell you. As for the people in the grandstand, they could see us coming, streaking right at them at their level. I'm sure their thrill was more of a crescendo of terror followed by instant relief. What a roller coaster ride—for that crowd and for me.

Now we disappeared again, letting the thrill subside, allowing the crowd to become absorbed once more in the other acrobatic goings-on. During this time, I moved from the front to the tail and got myself into a prone position in the tail gunner's slot. This was interesting, too. Back here in the tail, the yaw and pitch of the aircraft were more pronounced; the big tail, hanging way out here on the butt end of the aircraft, wagged irregularly around like the rear appendage of an uncertain but expectant puppy. We were out of sight for a good ten or fifteen minutes.

Back here, I was watching where we had been, lulled into relaxation by the hypnotic, undulating motion. As I watched, we descended again toward ground level, and I began to recognize some terrain features I had seen as we made our last departure over the crowd. This time, we were coming toward the base from the opposite direction, props saving the farmer the trouble of mowing his hayfield. Full power, max roto-rooter roar.

We were headed for the grandstand again, this time from the rear, and they would never hear us coming. We would burst low over their heads like a ground-zero thunderclap—hope there are enough ambulances to handle the cardiac cases. As we roared over them, it seemed like the tail dipped, and for an instant, I was looking that top row of gaping, panicked people right in the eye. Then we were gone.

Our little show was over. The precision aerobatic teams like the Red Arrows and the Thunderbirds had provided the oohs and ahhs. But we had delivered the microseconds of stark, raving terror. What an evil, satisfying feeling. This big ugly toad of an airplane had its moment.

It was time to resume the right seat for landing.

"Lower the undercarriage."

"Gear down"

"Landing position for flaps."

"Flaps 100%."

And that was how I ended up with one hour and thirty-five minutes of RAF flying time in my USAF pilot's logbook.

Old Shaky

By Charles F. Willard

I had submitted my paperwork requesting an "early out" from the Air Force after four years and over 2,000 hours of flying time. In late June 1965, my resignation from the Air Force was approved, along with those of several thousand other pilot officers. This was part of a government reallocation of funds to cover the political needs of the time.

Knowing I was about to leave his clutches, my squadron commander decided to wring every last flying hour out of me before I left the Air Force. Our C-130 departed Charleston Air Force base on the 26th of June 1965, return date indefinite. We flew back and forth all over Europe, the Middle East, and West Asia for nearly a month, resupplying our reconnaissance outposts and repositioning troops and equipment in response to aggressive feints by the Soviet Union. A plot of our routes overlaid on the map would have looked like a manic madman's cat's cradle, a mindless mass of crisscrossing strings.

Now, on the 20th of July, I found myself at Chateauroux Air Base in France, out of flying time for the month and the quarter. By Air Force regs, I was maxed out. Done. So, I had to leave my crew in mid-mission and find a way to "deadhead" home in time to get all the out-processing procedures and paperwork done before my separation date on the 2nd of August.

No direct military flights were leaving Chateauroux, France, for Charleston AFB, South Carolina, so I would have to zig-zag my way back home. If I could get to Evreux Air Base by tomorrow morning, a C-135 (the military cargo version of the Boeing 707 airliner) was leaving there at mid-day for Naval Air Station Norfolk in Virginia. From there, I might be able to hitch a ride to Charleston in a day or two.

As (bad) luck would have it, a C-124 Globemaster was scheduled to leave Chateauroux for Evreux that evening. It was a little over an hour's flying time in "Old Shaky."

The C-124, "Old Shaky," was the Air Force's largest transport plane until the jet-powered C-5 Galaxy came along in the early '70s. Old Shaky got her nickname because her cavernous interior was alive with all manner of gear and fittings that "shook, rattled and rolled" as her four

3,800 horsepower Pratt and Whitney 28-cylinder propeller-driven engines pulled her unlikely ponderosity, rumbling, shuddering, and vibrating through the air. Of course, she had other nicknames, too; among them, "the aluminum overcast" and "a million rivets flying in loose formation."

She was easy to make fun of. Parked on the ramp with her front clamshell doors agape and her hydraulically lowered ramps licking the ground, she reminded me of a hippo with its mouth wide open. From the side, squatting there, she resembled a haybarn with wings strapped on. If she could fly, then pigs certainly could. In fact, some of her pilots called her "the old sow." And she had this queer little radar dome that protruded from her proboscis like a blackhead pimple. One had the urge to take it between a thumb and two fingers and pop it.

She was designed after World War II as the Cold War ramped up. The sizes of most military equipment were growing like Topsy—larger tanks, bulldozers, guns, missiles—and the air transports available then were too small to haul them. That and the 1948-49 Berlin Airlift crisis made it abundantly clear that the Air Force needed a better heavy-lifting capability. Enter the C-124 Globemaster II—Old Shaky. She could swallow it all whole—the most oversized cargo the military had to offer.

She did not have an auspicious start to her career. Like many aircraft of her era, she had some early teething problems. A series of crashes starkly revealed her initial flight idiosyncrasies, control eccentricities, and design deficiencies. Over the years to come, Douglas built 447 C-124s. She ended up being an invaluable, if ponderous, fairly reliable workhorse. But, by the end of her 25-year operational life in 1974, one of every seven "Shakys" had met a violent end.

The worst of her early mishaps occurred on June 19, 1953, when C-124A, tail number 51-137, with a crew of seven, including three experienced pilots, departed Tachikawa Air Base, Japan, for Korea. Reliability was not her strong suit then, and when a C-124 taxied out, she was as likely to abort as fly. Her mission that day was to return 122 airmen from a five-day Tokyo R&R back to the front in Korea. She failed.

She took off into low cloud cover and visibility. Then, off the end of the runway, "in the soup," clawing for altitude, she lost an engine. Three minutes later, a red glow flared through the murk. She had crashed in a giant ball of fire. All aboard perished. Up to that time, it was—by far—the worst accident in aviation history. So, I was not delighted when I learned that "Old Shaky" would be my transport to Evreux.

It was a rainy evening with low, glowering clouds. I lugged my sodden self, along with my B-4 bag, crammed with a month's dirty clothes, up the steep, long ramp into her cavernous belly, feeling a bit like Jonah.

Except for me, she was empty—no other cargo. The loadmaster relieved me of my burden and threw a strap around it, securing it into a corner of the massive cargo floor, so the soggy suitcase would not become an overstuffed airborne projectile in the turbulence we were sure to encounter. I looked around at the interior.

It did seem like a haybarn. Inside here, she was over two stories high. There was a second deck, but that floor was folded lengthwise, the halves hinged up against her sides and fastened there near her ceiling. I glanced upward, almost expecting to see pigeons roosting along the top of the second purlin timber. Spanning the length of her roof ran a track with a couple of hoists suspended there. It reminded me of the hay hook track that ran along the ridgepole of our old barn at home.

I found a perch near the rear on one of those web-backed canvas seats that lined the sides. I chose to be far away from the line of the props, where the noise would be highest. Besides, if one of those seventeen-foot propellers over-sped and cut loose, it would slice right through that area like a cleaver through a pig hock. I found the lap belt and strapped in, wet and miserable. A cold front had come through, and it was a cool evening for mid-summer. It felt like an ice house in there.

The loadmaster secured himself even further back, next to the jump door. *What does he know that I don't?* The rest of the crew, schlepping their gear, climbed the long ladder to the flight deck, way upfront and way above. *They need an elevator in this enormous cowshed—or at least a dumb waiter*, I thought. It wasn't long before they were running the start-engines checklist.

I heard the whine of the starting motor as it began to turn the number three engine. It ground and moaned for a while, trying to get the thing to catch. There were some wheezes, a cough, a choke, a spit, and some irregular explosions as a few, then more of the spark plugs in those 28 cylinders fired the high-octane gas, and each piston reluctantly began to do its work. I could picture the flight engineer up there at his panel, working rods and levers back and forth, playing with the fuel-air mix, trying to get the damn thing going. Then, with a few more "rrrmps" and roars, the massive engine finally settled into a rollicking idle.

Then they were on to the next engine, and the agonizing process was repeated until all four reluctant radial juggernauts were rotating. *Geez*, I thought, *all we do in a C-130 is press a button, and the turbine winds smoothly up to speed.* We taxied out to the end of the runway and sat a while, the engines idling roughly—pocketa, pocketa, pocketa. I guess they were waiting for clearance to go.

Then, they ran the engines up to full power, checking everything before takeoff. Suddenly I could understand how "Old Shaky" got her name. Every fitting in that cavernous cargo bay came alive in a clamorous, frenzied dance. The gravelly howl of the engines and the clatter of all those accouterments rang, jangled, and roared like the Anvil Chorus performed at hyper-speed. Everything was shaking so much that I had double vision. Then we began the takeoff roll.

It seemed to last forever—like we were using three times as much runway as was available—and this was an empty airplane! As a C-130 pilot, I was always a nervous passenger. Someone else was flying the plane. Did they know their stuff? I was doubly worried, strapped to the wall of this would-be flying barn.

We made it off, somehow, into the clouds. Vibrating along here in the back of Carlsbad Cave, I thought the climb-out seemed agonizingly slow. Then the turbulence began. We were in the weather, not climbing above it as we would have in a C-130. The C-124 was not a pressurized airplane—how do you pressurize an airborne warehouse? We couldn't fly higher than the crew and passengers could breathe. Eight thousand feet was a typical cruising altitude, right where the worst weather lived. So, we jounced, jerked, and rattled along. It would have been a smoother ride in a springless old buckboard. Much quieter too.

I began to relax. The turbulence was tossing us around a little, the aircraft wallowing and bucking, but nothing I couldn't handle. I had never been sick in an airplane and wasn't about to start now. I had been close, once, on an ROTC "morale flight" in the back of a twin-engine, twin-boom C-119 "Flying Boxcar"—aptly named—through five hours of bucking bronc weather coming back to Syracuse from Valdosta, Georgia, one stormy night, but that was the last time I was even close to airsickness.

I relaxed too soon. I heard the sickening sound of an engine over-speed, one of the big motors winding up to a banshee shriek as the prop began to "run away." This happened when a propeller lost pitch lock and the blades flattened, relieving the engine of the load of pulling the airplane.

This was dangerous in all sorts of ways. First, the over-speeding pistons racing up and down at rates they were not designed to withstand could cause the engine to explode. Second, the prop, whirling faster and faster, might fly apart, sending blades into the fuselage and the fuel tanks. Third, a propeller pulls the aircraft by chewing into the air. This one was acting like a big barn door out there, yawing the plane into a skid and pushing it out of control. Unless the pilot could somehow stop the process, we were a goner.

But now I heard the engine begin to wind down. Whatever the emergency procedure was for this condition, it was working. The pilot had somehow regained control of the prop, gotten it to feather (blades to stand parallel to the slipstream), and had shut down the engine. That meant we were now flying on three engines. If this were a C-130, I wouldn't have worried—plenty of excess power on three engines—but this was "Old Shaky." The jury was out.

Soon, I observed an odd and distressing happening. The flight engineer trundled down the ladder from the flight deck, removed a panel in the side of the aircraft, and disappeared. *What?* I was dumbfounded. *Had he bailed out?* He wasn't wearing a parachute. I looked around to see if any chutes were stowed on board and saw none. *Oh, oh!*

But in a minute or two, he emerged back through the opening and buttoned up the panel. He must have seen my open-mouthed gawk. He came back to where I was and shouted in my ear. "We had an over-speed on number two and feathered the prop. The wing is so big that it has a hollow space in back of the engines. I crawled out to see if there was any visible fire, leakage, or damage. Looks like we're good. We should be landing at Evreux in half an hour or so."

"Th-thanks!"

We bumped along in the soup with no further problems. I felt us begin our descent for landing. We were yawing and pitching our way down through the turbulence, the rain, and clouds on final approach when I heard the dreaded sound again and felt us swerve almost sideways—another over-speed on one of the three remaining engines. My life passed in front of me. I thought of my pregnant wife and my child back home. What a way to end my Air Force career. Crashed in a flying trash hauler.

Somehow the pilot, flying on instruments, wrestling that big hippo down toward the runway through the rough and soupy weather, with a run-away prop, low and slow when an airplane is at its most vulnerable—somehow, that crew got the propeller under control, kept the aircraft flying, and got us safely on the ground with a decent "ker-thump" landing. Relief and sweat washed over me in that iceberg of a cargo bay. Not being part of the crew, I had no control over the outcome. That made it—without a doubt—the most frightening experience of all my time in the Air Force.

From Evreux, things went smoothly. I got my ride on the C-135 to Norfolk and found a Charleston C-130 from there to home, arriving with a week to go before my separation date. A scramble, but we got the furniture packed and sent to storage, the house sold, and all the out-processing done. We left Charleston for good on August 2. I was a civilian again. Now all I had to do was find a job and a place for us to live.

RVWG Meeting, January 6, 2022

Prompt #1: Write about the support you received in the service. How did it help you navigate the experience?

The Support I Took for Granted

By Charles F. Willard

I didn't realize it then—how much support I really got.

First from my family. They had seen loved ones go off to the service before, to World War II and Korea. Thankfully, all had returned mostly unharmed. An unusual blessing. Maybe they expected that's how it would always be—but I doubt it.

Six weeks after college graduation, when I packed my car to drive cross-country to pilot training, I got my mom's usual hug and kiss, and a handshake from my dad, but I don't remember any tears. Maybe they came later; I don't know. It was an adventure for me. There was no trepidation. I was excited to be on the road. "Be Careful." "Sure, Mom. Bye."

Mom saved my letters from those pilot training days, bound in a blue ribbon, tied with a bow. She must have treasured them to keep them all those years. Maybe she thought each one would be the last. I have them now, still tied with that blue bow, a memento of the depth of family care and encouragement that went with me when I left home.

When I graduated from pilot training, I asked my long (in both time and distance) girlfriend to fly cross-country and be there to pin on my wings. I never thought to ask my parents. I could only afford one ticket—it was $300 one way, I recall, all the money I had saved in a year from my $232.22 a month second lieutenant's salary—and no way my mom and dad to afford to come on their own. That was the beginning of girlfriend (later, wife) support.

Wife support began when I started my permanent assignment at Charleston Air Force Base in South Carolina. She signed on as a service wife rather than come with me as a girlfriend. Barely 21. Did she know what was in store?

Growing up on a dairy farm, I doubt she knew what being an officer's wife entailed—especially as an interloper Yankee in the deep south: White gloves, teas, name cards, elderly ladies from church dropping by to size her up—South Carolina traditions, strange to an Eastern New York country girl.

She got a job as an intensive care nurse at the Medical College of South Carolina Hospital, where she could barely understand the patients' and staff's Gichi and Gullah accents. Officers' Wives Club, USAF manual on proper decorum and expectations for officers' wives. How to be the Perfect Wife in furtherance of husband's Air Force career.

Phone calls in the middle of the night—secret missions to who knows where and how long. Wake the baby, drive husband to Base Ops at three a.m. Days go by—no word of where he is (or if he is). No cell phones, then, no Skype, no contact at all. Two weeks or maybe a month later, another phone call, probably also in the middle of the night. What will the message be? Thankfully, wake the baby, drive to the base, pick up the husband.

Air Force life becomes a routine. Running the household alone, rushing a fevered baby to the doctor. Sleepless nights with a crying child, nursing ear infections. When he recovers, chasing after the jet-propelled toddler all day long. Trying to shop with the hyperactive kid grabbing at everything in reach. Making decisions she shouldn't have to make on her own. Because they have to be made. Exhausting.

Long nights alone, wondering. Bonding with other young wives in the same boat, the consummate supporter's only support, the "service family."

Spending weeks responsible for it all, then back to second fiddle when the husband comes striding through the door.

All taken for granted. What was expected. The good Air Force wife. The way it was back then.

RVWG
VERITAS PER VIRTUTEM
MMXIV

A Requiem for Bob

Stories by the Rochester Veteran's Writing Group on the Passing of our Friend Bob Whelan

Forenote

Bob Whelan was a beloved member of the Rochester Veterans Writing Group and a great friend to all of us. His sudden absence leaves holes in our hearts as big as bomb craters.

On Saturday, November 7, 2020, we dedicated our monthly meeting to writing stories of our fondest memories of Bob and what his friendship has meant—and will always mean—to each of us. He touched each individual, as is plain from the stories that follow.

We use "prompts" as a catalyst for our monthly writings. But somehow, prompts or no, these stories would have come forth anyway. Bob had that great an impact on us.

Rest in peace, dear friend. You are a great example of why old soldiers should "...NOT Fade Away."

PROMPTS FOR RVWG SESSION SATURDAY, NOVEMBER 7, 2020: "BOB"

What better title could show up at the top of this page for November? I think you might see the appropriateness of these prompts, and for this one time, I choose not to number them:

1. What important advice would Mr. Bob Whelan give to a new Serviceperson, do you think?
2. What one piece of advice do you think Bob would give to *you* now?
3. How would Bob advise persons who would be *leaving* the Service?
4. What changed in you because of Bob's sudden passing?
5. What did I learn about or because of Bob?
6. Do you recall the most recent thing you & Bob talked about? For example, did you talk about future ideas? About what you and he wanted to do next?
7. Do you know if Bob opined about what more he had in mind to write?
8. Maybe you very understandably do not want to write about Bob out of respect for the man and his life. Please do write anything that moves you.

Memories of Bob Whelan
By Sue Spitulnik

Nov. 7, 2020

I think Bob would tell a new service person to be more aware of the rules and attempt to follow them better. It might keep one out of trouble a bit more but wouldn't be as much "fun."

The last subject Bob talked about in the group was relationships with your parents and what kind of job you had done as one. His story in our book about his Aunt Jackie he felt was an example of a good mother. He thought now as oldsters we have the opportunity to set a good example of what true caring can be like. He felt like a great Dad to me.

I think Bob would advise someone who is leaving the service to stay connected to other veterans and to be involved in a group that understands how ingrained the military way of life is. I think he would share his belief there are many ways to serve at all times -- be friendly and make people smile or laugh is the main way.

What changed in me when I heard of Bob's passing? I became angry! Bob told his daughter Patty, and friend Steve, that he was in a good place and ready to go. Well, I wasn't ready to have him go! I hate that the death of loved ones leaves such holes in our hearts.

Bob taught me about service, giving, friendship, the love of life, and to set a good example by going the extra mile. The last time I talked to him via telephone, when he was in the hospital, he was concerned about getting the prompts ready for our November meeting. His mind was still focused on his responsibilities to others. After the meeting last month, while I was taking him home, he said he was excited about the next book project. He felt it would bring the members closer by working on a project together, and it was always good to have a goal to reach.

I know Bob had material ready to go into his own memoir. He wanted to find someone to do the formatting for it. I don't know if Patty might be able to put her hands on the information so it can become a book someday.

My car did a little skip on the way here today when it went past Bob's street because I was going to be the privileged one who gave him a ride to the Rochester Veterans Writing Group.

BOB WHELAN – WORLD WAR II VET, STORYTELLER, AND FRIEND

By Tim Hansen

09 November 2020

I can still see him staring at me from across the conference table with that mischievous glint in his glacial blue eyes and a conniving grin on his face. It's hard to believe he is gone, but his words echo in my mind.

Though our narrative styles differed— he much more philosophical— nevertheless, I always listened to him share the glimpses of his life and the wisdom that comes from many lessons learned along the way.

He grew up in Buffalo, N.Y., during the Depression. Life for most Americans back then was a constant struggle to put bread on the table, and maybe that's where Bob became a bit of a scrapper. When the war came, he answered the call, enlisting in the Army. He served as an infantryman in the 59th Armored Infantry Battalion of the 13th Armored Division.

Through Bob, we learned of the last bitter fighting of the war in Europe as Allied forces enveloped Wehrmacht units in the Ruhr valley, an operation referred to as the "Ruhr Pocket." The year was 1945. Casualties were heavy, and life was a bit grim, with death beckoning at every street corner, abandoned building, and home. Despite those dark days of fighting and dying, Bob spoke with great compassion, never uttering a bitter word about what he endured. He often recalled life's lighter moments with officers saying one thing and noncommissioned officers doing another or facing disciplinary action for missing a curfew.

In all of his stories, I heard that recurring thread of deep regard for all of humanity. His war narratives are priceless to me now.

And, he made me wonder whether life has moments of serendipity. Bob once recounted the time he and a friend were leaving his church on one wintery Sunday morning and saw a woman slip on a patch of ice. Ever the man of action, Bob dashed over to help her back up. They talked for a bit, and Bob asked her where she was from. When she told him of a small town in the Ruhr valley, Bob knew it immediately. His battalion fought and secured that town. The woman was only five years old at the time. Bob made a new friend that day.

Bob had a comment for everything, and sometimes I had to think for a moment about his words before realizing his point. In everything, he advocated unity over discord, prejudice, and ignorance. He took the time to listen deeply to the other writers in the group and always had an uplifting word to say to them. His friendship was infectious.

I admired Bob for his candidness of thought, his courage to speak up and be heard, and his beguiling smile before moving on to the next story. I realize now that is what I need to do in my own life...say something that makes people stop and process what you just said, wait a moment, and then smile.

Rochester Veterans Writing Group, 11/7/2020

By John Steele

Bob Whelan died this past week. Hearing the news has been very saddening but has elicited so many thoughts running all over the gamut that I can't help but smile a little bit. Some of the prompts for our writing group this week deal with what Bob would say. Well, he'd for sure encourage us to share out stories. And, of course, he would be positive. But beyond the predictable, the rest would surely fall under the realm of improbable but very true. I loved hearing his anecdotes. Here is this kind old gentleman, very sweet and humble. Then he'd share some of his antics and shenanigans from the past, and you realize that this kind, sweet old man was a hellion. Not malicious, but definitely a little mischievous.

Bob reminds me of my own Dad. My Dad was a World War II and Korean War veteran, a bit of a hell-raiser, impish, always smiling. My Dad died nine years ago. My Dad never shared many accounts of his time in the service. I, at times got little snippets, but never the full yarn. But when Bob started divulging some of his escapades, I could picture my Dad there with him. My Dad proudly boasted that he made Sergeant three times. However, he also made Corporal four times because each time he made Sergeant, he would eventually do something and get himself busted back down to Corporal. Tell me that doesn't sound like Bob.

I don't write as much as I'd like to or as much as I should. I want my daughter to know some stories about her Dad and be able to share them with her kids. I want my siblings and nieces and nephews to be able to see another side of me. I know Bob would want that too.

BOB

By Suzanne Dianetti

I was at work this morning and went into the Salute Room at Warrior Salute Veteran Services. I walked—I swear—into that room with Bob Whelan.

I remember last month that he spied the comfy chair and made a bee-line for it. The comfy chair remains empty today.

It's that smile, that hello, that amazing eye contact he made no matter what was going on or who was speaking and sharing. Mostly, I'll miss his presence and his laugh...I just miss Bob.

I always knew that somewhere along the story he had written that day he'd glance up with those devilish eyes and smile and I knew it was coming – some hysterical recall of an event that lent itself to mischief and a belly laugh.

His amazing knowledge and craftsmanship of words – his always reminding us of the laughter in the hardest of times – sausages, stealing an officer's car, and even a captured photo of it – but always laughed and smiled with his beautiful eyes.

What changes now with his passing is the void he has left. That voice that resonated with wisdom for one, but I miss a friend I wanted so much more to know – and that October hug that I wanted but COVID19 stole from me, from us all. I can hear his voice still.

Bob always listened deeply. It was important to him to listen deeply, and even, several times, I remember his comments on my writing in Group, applauding what I offered even when there were tears and vulnerability on my part – I felt his spirit and his hand on my shoulder today as I wrote this. Any offering from anyone Bob found of deep value and made it clear why it was valuable. He dug in with questions as he absorbed our stories, and Bob loved this family of writers, just as we did/do him.

I think now Bob would say DON'T STOP! Keep writing, keep smiling, laughing, and finding the goodness and peace in what could be chaos. Roger that, Bob.

He always asked after my son, Mike, asking about his whereabouts and safety, and it wasn't half-assed, it was genuine because that was and is Bob.

It's probably not appropriate to cross branches, but to Bob today, I say Semper Fidelis, dear Bob, Always Faithful.

Bob, So Glad We Knew Ye

By Charles F. Willard

Most veterans are survivors.

Yes, I know. Some of those veterans survived only to succumb later to the unspeakable damage inflicted on their minds and bodies by the horrors and revulsions they suffered. And many survivors would not speak–could not ever speak–of those experiences. A kind of buttress against the awful ghosts of their relentless, always resurgent memories.

But I imagine that one of the reasons we call World War II veterans the "Greatest Generation" is that they experienced so much, for so long, at such young ages, and, for the most part, went on to live long and fulfilling lives, raising families, starting businesses, giving their time, energy and intelligence to countless worthy causes.

That brings us to our World War II veteran friend, Bob Whelan.

He came into our Rochester Veterans Writing Group one day, a man in his early nineties. He never lacked for something to say. At first, some of us young whippersnappers in our eighties, seventies and sixties thought he was an old windbag.... And then we started to listen. Came the dawn! Here was a man of some considerable substance. He didn't talk about the horrors. His war stories were soaked in the human spirit, touched with humor, slathered in self-deprecation, all wrapped in a warm blanket of compassion and forgiveness. He had a depth of understanding and empathy for the human experience that seemed to escape many of us. He was a philosopher. The philosophical eludes a lot of us–too quick to jump, too quick to judge.

Bob often painted a comical picture of his soldier days. He was just a young kid when he began his service–as so many were. Remember your level of maturity when you were 18? I remember mine. A smug, wise-ass know-it-all, flouting authority with a passive-aggressive smirk, screwing up in any way I thought I could get away with (and sometimes did). According to Bob, that was Bob. And you could believe it, the way he told his stories.

Went in a Private, came out a Private. Mugging for the Movietone newsreel camera in front of the General's pomp-and-circumstance surrender ceremony. Too bad he could never find that film. Stealing sausages in a captured German town–only to meet up with the victim of his theft years later in a Buffalo sausage shop. Getting the "shit detail" of traipsing a gaggle of German prisoners back to the rear and wondering, "Why me?" He really did know "Why him." So many stories. None of them of the horrors. You wonder...

Bob, the later man, fired in that growing-up crucible where every young man served, and many of them died. A far cry from today. That difference is one of the things they, Bob among them, fought for, I guess. A better, richer, more genteel life where the very few fight for the very many while those "many" stay home, unaffected, safe, and cozy. I don't think Bob was bothered by this. He only wanted the rest of us to understand.

Bob, the later man: No one could believe he was once that wise-ass kid. Bob, the later man: Gentle, thoughtful, compassionate, forgiving, full of humor, life, and love.

Bob, so glad we knew ye.

What did I learn from Bob? To listen a little better, to judge a little slower. To maybe look a little deeper, think a little longer, and love a little stronger.

Bob, so glad we knew ye.

Excerpts from Evelyn Farrell's note to Sue Spitulnik on 11/8/2020

...So sad that Bob is gone...

Thankfully we got to talk to him at your place and I told him how much joy he brought to us. I told him he always said at the end of his talk ...forgiveness and love are what we need to do.... I told him as a dear friend I love him!! Because he lived what he believed... and we hugged with tears in our eyes ...

Chuck was in awe of him. He was a special guy!!!

We all are blessed to have known him...

Remembering Bob

By Joe Mele

I can still hear Bob enthusiastically and with a voice that carries well, relating stories that exemplify a theme that runs through his writing "The healing power of time."

We see his insightful perspective, born of years, experience, hard-earned wisdom, and love.

We learn that time, when tempered with forgiveness, can heal wounds and reconcile relationships.

Bob was a few years younger than my father, but they shared the experience of fighting with a WWII armored division in the European Theater of War.

My father and I did not communicate well, and we did not come to an understanding before he passed away in 1982. I learned that Bob had a troubled relationship with his father, but they eventually reconciled.

When my mom passed away in 2012, we found hundreds of letters—many of them from my father during WWII. Through these letters, I've come to know and understand my father better than I did when he was alive. With Bob's example as guidance, I've come to terms of a one-sided sort with my father.

God bless you, Bob, and Thank You.

BOB WHELAN

By Lee Nelson

I took a shot on a new club.
That's when I met him.
He was a striking man
sharp
intelligent
impressively vigor.
I later learned he was over ninety.
He was doing great for late seventies
and a thought popped into my head.
For all his battles, this man has won another:
The Battle of Aging Gracefully.

The club proved fruitful to my soul
and on an awesome picture splendid day
the man offered his potato chips to me.
I ate some chips and later talked about my mother
a story about her and me and the military.
The story tickled him.
He laughed.
He looked at me and laughed.

All I ever wanted to hear

was what he had to say.
As soon as I'd met him
I knew this absolutely.
Bob Whelan had humility
and no regard
for his honor or wisdom,
but what did I know of this
diminished man chewing chips?
Very little, but I felt very little.
I felt as little as anyone would
when a giant gentle soul
breezily extends regard.
I got what I got
and it's no sappy overstatement
that I am changed.

And now
I'm more a stranger than any
among a world of close people
who reckon his Taps.

I do believe I met a master
a master of life
and the human condition
and the grace of knowing one
makes a man a better man.

The long traveling meteor
passed me with a flash.
It was Floyd's fleeting glimpse
what he gave me
but I feel touched.
Words fall flat to any majesty
so touched will have to do.
I feel touched
dear reader.
I feel a finishing
and decisive
touch.

TRIBUTE TO BOB WHELAN
By Cindy Blair (USN Veteran 1990-2002)

When loved ones pass away it is never easy to accept that they are gone. People who know you may say things like, "They're in a better place.", "They aren't suffering anymore.", or "I feel your pain." These remarks don't seem to make you feel any better. Perhaps it might be better to not say anything at all. I say, "I am here for you."

What does seem to help is when those who also knew the deceased can share loving memories of that person; as evidenced in this Rochester Veterans Writing Group. Most of us have heard about the different stages of grief. Sometimes I think people don't need to go through all the stages before healing begins. I can't imagine any of you getting angry at Bob for leaving you; because it sounds like he would have stayed a while longer if he could have. I regret not meeting him in person, but his memory lives on in you all.

Suffice it to say that grief is not static; it is fluid. It comes in waves of sadness and fond memories. Fortunately, when that beloved person is an artist, in Bob's case a writer, their work is a lasting legacy for strangers and loved ones alike.

Due to sheltering-in and isolation caused by the pandemic, a different kind of grief and loss is felt by many. I miss singing in a group the most. I had just joined an a cappella chorus in January 2020. We were scheduled to perform in a regional competition on April 1st. I spent money on costumes, stage make-up, entry fees and other costs associated with competing; only to have that thwarted by the pandemic. Now our rehearsals are held on virtual platforms. We all sing at the same time but we are all on mute. Those sweet melodies and harmonies are falling on deaf ears. The foreseeable future for our singing group is bleak. There is talk of holding competitions by video submissions but even those plans are on hold.

Lessons for surviving this loss have been, for me, to continue honing my vocal skills and start other worthwhile projects, such as this writing group to fill in the gaps. I am so pleased and proud to be a part of the Rochester Veterans Writing Group.

"Figuring Things Out"

By Vaughn Stelzenmuller

Bob was a true *figurer-outer*. Maybe that's *figure-outerer*. This is the ability to figure things out that come along in life, or figure out how to solve or fix a problem or situation that pops up, or figure out how to *get out* of something. His ability to do so was a gift, and one that served him well in life.

This talent led him on to grow his second gift: a very wide *sense of gratitude*.

Then Bob fell.

Right away thereafter, what happened reinforced a conclusion that I had mostly *figured out* many years past.

Now, I have known many older folks, all of you likely have too, who, late in life got switched into a much different situation, suddenly. I personally had a mother, a couple of mothers-in-law, a father-in-law, some aunts and uncles, and assorted friends and friends of the family who had found themselves in the same spot. These good people got suddenly put into different place, by equally good people with the best of intentions—really, with the most loving of intentions. Then they fell, or something disabling happened to them. Then things turned worse.

Bob fell. Then he had the worst of Hobson's Choices thrust at him and his good family: to go to a hospital. I truly do not think he could "figure things out" this time. And then, *things* turned worse.

It was then that his second gift kicked in. From what I have been told, he died at some kind of peace, which could only come from that deeply-understood gratitude which was now his hard-acquired nature. *Heck*, you can see both of these gifts in the stories he told. Both gifts!

His memory very starkly pointed out to me the warning about late-life personal emergencies. The kind of emergency that I have seen happen many times for many people I knew.

At the same time, I know he would have told me that we have to be humble about being able to "figure things out," because sometimes we can't. Sometimes, we just can't. I also feel him pointing his finger at me to get busy maturing that *feeling of gratitude*. This, he tells me is something that everyone *can* do.

Hey, even as I gladly give him the highest epitaph—SOLDIER—I would remind everyone about his casual advice:

"You have to take risks in life, and sometimes you get caught"

Well and humerously said, Bob, and maybe good advice to *new* persons entering a life of Service.

Made in USA - Kendallville, IN
38547_9780984179220
06.20.2023 1309